KU-203-931

Fired by Steam

Fired by Steam

GEOFFREY WHEELER

JOHN MURRAY

Sources of Colour Plates

1 Robert Frankcom Esq/National Railway Museum, York
2 Geoffrey Wheeler
3 The Hon. William McAlpine
4 Geoffrey Wheeler
5 Geoffrey Wheeler
6 Geoffrey Wheeler
7 G. Gittins Esq
8 Geoffrey Wheeler
9 The Hon. William McAlpine
10 Geoffrey Wheeler
11 Tony Dyer Esq
12 Geoffrey Wheeler
13 Geoffrey Wheeler
14 Geoffrey Wheeler
15 The Hon. William McAlpine/Solomon & Whitehead Ltd
16 The Hon. William McAlpine
17 Geoffrey Wheeler
18 G. Gittins Esq
19 Tony Dyer Esq
20 D. A. G. Marshall Esq
21 The Hon. William McAlpine/Solomon & Whitehead Ltd
22 Geoffrey Wheeler Esq
23 Geoffrey Wheeler Esq
24 Susan Wheeler

FRONTISPIECE: Great Western 'Bulldog' 3416 *John W. Wilson* and
'Star' class 4034 *Queen Adelaide* working as one as they breast
Hemerdon bank on the haul eastward from Plymouth in the early
1930s (*R. C. Riley collection*)

© Geoffrey Wheeler 1987

First published 1987
by John Murray (Publishers) Ltd
50 Albemarle Street, London W1X 4BD

All rights reserved
Unauthorised duplication
contravenes applicable laws

Typeset by Keyspools Ltd
Printed and bound in Great Britain
by Balding & Mansell Ltd, Wisbech

British Library CIP data
Wheeler, Geoffrey, 1929–
    Fired by steam.
    1. Locomotives—Great Britain—
    History
    I. Title
    625.2'61'0941   TJ603.4.G7

ISBN 0–7195–4383–5

To Enid

Who came train spotting and stayed
– and who always says 'Thank you'
to engine drivers

Beside those spires so spick and span
  Against an unencumbered sky
The old Great Western Railway ran
  When someone different was I.

—from 'Distant View of a Provincial Town',
  Continual Dew (1937) by John Betjeman

Contents

Acknowledgements

I am very grateful to all those people who, like me, love to watch trains — especially steam trains — go by and who have, over the years, written to me in their hundreds, indeed thousands, from all over the world in respect of my steam-driven illustrations; to Duncan McAra of John Murray, whose idea this book was in the first place and who didn't chivvy me too much when I started to run behind schedule; to my wife Enid and our family, Dave, Jon and Sue who, I believe, really did try to tolerate me when the written word wouldn't come — but the spoken word did; to Dick Riley who found most of the photos I wanted, including some 'impossibles'; to Kenneth Bound, Editor of *Mayfair*, who has supported my work in Christmas issues in recent years; to Peter Kelly, lately Managing Editor of *Steam Railway* magazine, who 'drew me out' at my fireside one winter's day and who rekindled old half-forgotten enthusiasms for the Great Western, then wrote so kindly about me; to Reggie Collins of the GWR, my mentor and friend 'who knew everyone' and to whom I never did say thank you but to whom I do so now; to the late John Solomon of Solomon & Whitehead (Guild Prints) Ltd; and last but not least, to Messrs Boville Wright of Uxbridge who have kept me supplied with artists' materials for more years than I care to remember, even when my bills were not paid on time.

And, of course, the Great Western Railway itself.

I thank you all.

It was one of those rare and beautiful afternoons in February, when all around the day seemed to be standing still – crisp and golden with not a breath of wind. Only England could produce a day such as this in mid-winter, I thought, as I stood on Beaconsfield station waiting for one of those 'new-fangled' diesel trains to take me up to London. In those days in the mid-1960s, Beaconsfield still boasted four tracks: two through roads and two platform roads. As I stood there, the 'sticks' (signals) came off for a 'down fast' on the through road – a Birmingham express. I watched its smoke in the distance as it came nearer, breasting up the rise from Seer Green then out from under the far bridge, to come thrusting through the middle road, striding along with that thumping roar so typical of a Great Western four-cylinder engine working hard – 6026 *King John*, with all the majesty a 'King' could muster, resplendent in green (I even forgave its British Railways emblem that day), and behind it a long heavy train of immaculate chocolate-and-cream coaches, all perfectly matched. I waved to the driver. He lifted his hand from the cab rail a few inches in return. Moments later, it had all gone. The only reminder was the 'clunk' of the signal going back, in that quiet stillness which comes after a train has passed, and an almost-stationary trail of smoke left hovering in a dead straight line along the length of the station and which seemed reluctant to go away. As it slowly dispersed my diesel came in.

Somehow, I thought, I had to try and recapture something of that moment and make it live on. Start now, before steam was gone for good. But how? I wasn't much good at painting pictures, not in the normally accepted sense – the respectable way. I had my drawing board, though, and my pencil. Then there was my airbrush – and what about all those works' drawings of locomotives in the drawer that I had collected over the years? And had I not lived and breathed steam, day in, day out, for as long as I could remember?

My earliest recollection of anything to do with railways is, appropriately enough, a picture book. I can see it now – one of those books with thick cardboard pages, each with a coloured picture of a train on its shiny surface. Together my father and I would go through its pages again and again at the fireside in our cottage at Penn until I knew every detail of its contents and

Regal splendour: 6026 *King John*, fresh from the paintshop at Swindon in June 1952, looking much as I saw it when it aroused new interests on that fine February day over a decade later, only here it displays the lion-and-wheel on its tender and not the later, more ornate BR emblem, the one I forgave (*R. C. Riley*)

The 'cigarette card engine': 5005 *Manorbier Castle* at Leamington Spa in 1936, complete with bullet nose and streamlined fairings. These trappings were removed bit-by-bit over the next few years but the wedge-fronted cab and long splashers lingered on until 1947 (*J. A. G. H. Coltas*)

could recite the name of each picture by heart, long before I had mastered the art of reading them. To this day, I can recall those trains and conjure up their images in my mind. They were the 'set pieces' of their day – the Flying Scotsman leaving King's Cross, a red Royal Scot wending its way through heather-clad hills, and a Southern Railway express at speed alongside the racetrack at Brooklands. Most vividly though, I can recall one picture – an impressive, low-down, threequarter front view of a train at speed over water troughs. All green and steam with a suggestion of brown and cream coaches half hidden by spray, it seemed to come bursting out of its cardboard page. 'That', my father would say each time we looked at it, 'is the fastest train in the world – the "Cheltenham Flyer".'

It was, I suppose, a neighbour who really set the ball rolling, though quite unwittingly on his part. Fred was a friendly, gruff old character who lived in the cottage next door, and who would, as regular as clockwork, come rolling home each Sunday afternoon – with a degree of cheer – soon after two o'clock when the local pub, the Dog, had closed its doors on him. This time, he didn't sing his way down the garden path but, instead, called me to the hedge where a hand appeared through the prickles and handed me a bundle of cigarette cards. This unexpected gift could have well been cards of film stars or cricketers, but to my utter delight they turned out to be an almost complete set of 'Locomotives of the World', and there amongst them I soon found one which immediately became my favourite – the Great Western's new bullet-nosed streamliner, *Manorbier Castle*.

A rare and extra special treat the following Christmas in 1936 was to be taken for the first time to see a circus at the Agricultural Hall at Islington. It was a truly mammoth affair, and I remember watching enthralled as lions and tigers were made to behave themselves in the ring and a troupe of white-clad performers carried out amazing balancing feats on glittering silver bicycles on the high wire. I held my breath and a woman near me shrieked as a bespangled girl suddenly plunged earthwards from the flying trapeze, to be brought up only inches from the sawdust by an almost invisible strand of steel wire attached to her ankle. But there was yet another thrill in store for me that night, and one which completely eclipsed the circus. Coming back late on the Tube to Paddington station, all-of-a-rush to catch the last train home, we came hurrying up the escalator and were dashing across the station when, to my amazement, I saw it. There was no mistaking that bullet nose – it was *Manorbier Castle*. Naturally, I stopped. The rest of the family didn't. Frantic parents returned, grabbed me, and bundled me into an already moving train where, amongst much shouting, slamming of doors and blowing of whistles, I desperately tried to explain about *Manorbier Castle – my Manorbier Castle*! Great Western engines weren't just pictures. They were real!

In those days, going anywhere by train was a rare treat and something of an adventure; and although I cannot remember my very first train journey I do recall that whenever we were taken to London – always by train – my brother and I had to wear name-and-address labels in our lapels so that we could be returned home like parcels, should we become lost. I recall, too, as a very small boy, hanging grimly on to the leather strap that raised and lowered the carriage window and finding, to my great delight, the letters GWR embossed on it. I knew the train we were on couldn't be the 'Cheltenham Flyer' as it kept stopping so I consoled myself with the knowledge that this must surely be the fastest stopping train in the world.

I grew up in the 1930s when steam was still very much in evidence in the countryside, and not only on the railways. The occasional steam waggon (always referred to as a '*Foden*', no matter what its origin) still went hissing its way through the village, and threshing tackle came to the farms steam-hauled by traction engines each winter, to be followed by its supporters' club, sometimes for miles, and which always seemed to include me. A portable steam engine with a tall slender chimney leisurely worked a sawmill on the other side of the woods, steam rollers left their marks on roads, and the fair with its steam-driven roundabouts arrived at the village green each year in the charge of the most glorious array of crimson-and-gold showmen's traction engines – a feast of colour and atmosphere to delight the heart of any small boy, particularly as it sprang up like magic, almost overnight, right opposite my school without fail every September, and where, for two whole days and nights, it ripped apart the calm of the village, and where local men who had born grudges throughout the year, settled them with their fists in the glare of the arclights and naptha flares.

A familiar scene from boyhood and one little changed when this photo was taken nearly twenty years later in June 1953: an ex-LNER (originally Great Central) A5 tank, 69814, skirting my native Chiltern Hills near High Wycombe on an up stopping-at-all-stations train to Marylebone (*Brian Morrison*)

Another Chilterns scene: the 9.10 Paddington-Birkenhead (one of the two-hour Birmingham expresses) between West Wycombe and Saunderton on a misty morning in September 1960, with double-chimneyed 6013 *King Henry VIII* up front (*M. Pope*)

At a very early age, Santa Claus saw fit to bring me, and a number of friends, small brass German-made toy steam engines which we were allowed to operate unsupervised, and whose red flywheels were hardly allowed to come to rest throughout the whole of the Christmas holidays. Pools of burning methylated spirits soon made holes in carpets and took away the shine on table tops, so we were very quickly banished to the garden shed which, miraculously, survived its baptism of fire. Later, and for the princely sum of tuppence, I bought a small steam locomotive from a village jumble sale. It was a pretty little thing – also German-made – a replica of a Great Northern engine, and one which, I am sure, would be considered a very collectable item these days. It boasted such sophistication as a whistle and a beautifully made slip-eccentric valve gear operating its two small cylinders, but lacked a burner, so I bought penny candles which, broken into stubs, were lighted and forced up between the frames under the boiler to raise steam. All went well until this method of firing melted the solder around its tiny pipes, so I sold it to another boy for five shillings – and promptly went out and bought another steam engine, a vertical-boilered 'coffee-pot' machine (again of German origin) which I had seen standing in splendid red-and-brass isolation among the broomheads and paintpots in the window of a local ironmonger's shop.

Hornby trains, the lovely old tinplate variety in gauge 'O' (pronounced 'gorge nought', I remember), came into my life, and that of my friends, on the way to school one morning when we discovered that profusely illustrated catalogues of Meccano's superb products could be obtained, in anticipation of Christmas, completely free of charge in the post office just around the corner from the school. Word soon got about, and by lunchtime the entire stock of catalogues had been snatched up, to be drooled over in the playground or secreted within books and browsed over during lessons. Every boy (and a good many girls, too) in the school must have had a copy. The following year, I recall, the price of a catalogue became a penny, which reduced its circulation but not my enthusiasm for Hornby trains. Equally enthusiastic was my bosom pal Norman, and although some time was to pass before we were to be blessed with Hornby train sets we both received them together one Christmas – not the expensive scale models for which

Hornby were so renowned, but simple clockwork sets, which didn't deter us one bit as each of our engines proudly carried the legend HORNBY on its shiny black smokebox, which was all that mattered. I did envy Norman a little though. My engine was a black LNER whereas his was green – a Great Western.

As youngsters, we enjoyed a complete freedom and the rough-and-tumble life of country boys. We fought, climbed trees, trespassed at will, helped ourselves to the farmers' turnips, and generally ranged far and wide on foot over the wild woods and fields of the Chiltern countryside, still unspoilt by developers in those days. So many times, these expeditions would end up at the railway (the Great Western/Great Central Joint Line) some three miles away, which drew us like a magnet and where, any differences forgotten, we would linger for hours, perched on the strands of the post-and-wire fence, simply watching trains go by. It wasn't long before we had worked out between ourselves the routine of this railway, with its black tank engines (later identified as ex-Great Central Robinson A5s) on suburban trains, and the heavy Great Western 'two-hour' Birmingham expresses, usually headed by a 'King' – and on one never-to-be-forgotten occasion by *King George V* itself, still basking in its transatlantic glory and proudly sporting its gleaming brass bell, to be sent speeding on its way by a knot of grubby little boys who waved sticks and cheered for all they were worth.

This railway line had another, and probably wider, influence on the village where I lived than that of a means of transport as it was our weather forecaster, and one which never failed. When you heard the trains loud and clear running along the valley between Beaconsfield and High Wycombe (with a pause as they passed through Whitehouse tunnel) it was a sure sign of rain, and many were the times I saw my mother gather in the washing from the line on a Monday because she could hear the trains, and sure enough, within the hour it would be raining. The railway's influence came right up to our doorsteps, too, in the shape of a big chocolate-and-cream delivery van from the station – a Thorneycroft – a flat-fronted vehicle which seemed always to give the impression (to me, anyway) that it had run head-on into a brick wall, so abrupt was the flatness of its front end.

A satisfying reward at the end of a 15-mile cycle ride from home: one of the streamlined 'Duchesses' of the 'Princess Coronation' class, 6227 *Duchess of Devonshire*, in wartime plain black livery on the LMS main line at Berkhamsted in July 1944 (*H. C. Casserley*)

95 *Robert H. Selbie*, one of the four Metropolitan Railway 'namer tanks', a regular at Great Missenden on the Metropolitan/Great Central line through which it and others ran after taking over from electric locomotives which brought the trains from the City and Baker Street as far as Rickmansworth station (*N. Shepherd, R. C. Riley collection*)

The love of steam, they say, is like a disease – one which settles in your system and stays put. Let it get a firm hold and no power on earth will ever make it go away. There is no known cure, so you learn to live with it. It can lie dormant for years but it's always there, and then, like Mount Etna, it can erupt quite suddenly – without warning – then go back to bubbling away in the background. There was no hereditary reason why I should have caught this affliction. My father was a carpenter and his father a builder, as was his father before him, and so the family had been for generations in our village. On my mother's side, Grampy Miles (Toby) made chairs, as so many did in High Wycombe, and would take them to London piled high on his horse-drawn van, where he would cry his wares around the streets and not come home until the last stick had been sold. There was, however, Uncle Ted who, my dad told me, at one time drove a steam tractor and who, more than once, tied down its safety valve (to the consternation of his brothers – who scattered) in order to gain extra steam to shift a stubborn load out of the brickyard, and not so many years ago a long-lost cousin turned up at my house and announced that he was an engine driver on the LMS, so perhaps my roots in steam were not quite so remote after all.

My brother had been bitten by the 'steam bug', though to a much lesser degree than I had been. Bill, two years my senior, went to the Technical School in Wycombe, which backed on to the railway line near the station, hence his interest (and that of his classmates) in trains. The term 'train spotting' had yet to be coined in those days. One simply went 'name collecting' or, as some called it, 'copping'. Hanging around on High Wycombe station one Sunday afternoon when rail traffic was at an all-time low, Bill and I discovered that, for a few pence, we could reach Maidenhead by train, and where, almost as soon as we had set foot on the platform, 5042 *Winchester Castle* came storming through with a tremendous roar, seemingly only inches from where we stood in our efforts to record its name – the first of hundreds of Great Western locomotive names I was to collect.

Here was my new territory to be explored – a positive paradise, where trains came and went every few minutes, with engine after glorious engine steaming through, some fast, some slow, and each having a name (or just a brass-bound number) worthy of being recorded in grubby notebooks, to be copied out in 'best books' (usually sixpenny cash books) at the end of the day – in handwriting far neater than any we would think of using in school.

Bill's interest waned. Mine increased, and soon I had become a regular visitor to this wonderland full of trains, the Great Western main line, to fill notebook after grubby notebook with names and numbers. A bike became my magic carpet and with a Government-surplus haversack containing notebooks and sandwiches I would be away all day, thoroughly enjoying every moment for the cost of nothing more than a bottle of pop, now extending my lineside activities to such places as the LMS main line at Berkhamsted for things like 'Royal Scots' and streamlined 'Duchesses', and the Metropolitan/Great Central line at Missenden for LNER Pacifics and Met. 'namer tanks'. But always I would return to the Great Western main line where I felt my loyalties lay and where I was happiest, especially as I had now found an ideal spot half-a-mile or so west of the station where a bank beside the railway gave a commanding view of the line and all the signals ('pegs') heralding the approach of trains.

Now the Second World War was in full swing, and around the age of twelve an early morning paper round coupled with 'lending a hand on the land' at a nursery garden (turned over to food production) brought me seemingly untold wealth, so I spread my wings further afield, travelling all the way to Swindon on my own where I 'jumped' a fast train to Paddington headed by 4083 *Abbotsbury Castle* just to get the 'feel' of the old 'Cheltenham Flyer's' record run. I knew my ticket didn't cover my going all the way to Paddington then coming back down to High Wycombe, but a kindly ticket collector, who took pity on me, advised me to dash through the barrier just before the Maidenhead train was about to leave from Paddington and simply wave my ticket and shout 'Wycombe'. I did. It worked. Then I caught a train from Maidenhead where my ticket was valid and came back to Wycombe round the branch line. *Abbotsbury Castle* did me proud that day, and although it broke no records we had a fine non-stop run right up to the buffer-stops at Paddington.

Another time I wandered down to Basingstoke to savour the Southern Railway, and you really did wander in those days, down the leisurely branch line from Reading where, every time the train stopped, everyone

'I was a boy myself, once,' remarked the friendly ticket collector, who let me ride non-stop to Paddington on a 'wrong ticket' behind 4083 *Abbotsbury Castle*, seen here passing Teignmouth Quay on the down Devonian in July 1958 (*R. C. Riley*)

My youthful wanderings took me down to Basingstoke where I first made the acquaintance of the Southern Railway. Soon afterwards, SR 750 *Morgan le Fay* (seen here at Basingstoke as BR 30750) was to visit my locality, along with other 'King Arthurs', on the GW/GC Joint Line when they worked on goods as a wartime expediency (*R. C. Riley*)

seemed to know everyone else and the Great Western stopped for a chat. On the way back I stopped off at Reading for a while, long enough to find 'Bulldogs' in the sheds there, though I had to travel to Didcot another day to get my first 'Duke', 3283 *Comet*. Of all the GWR engines I had got to know, I think I liked the 'Dukes' and 'Bulldogs' best. I 'found' 3265 *Tre Pol and Pen* on home territory, on light goods, standing at the 'pegs' as I came over the station bridge at Wycombe on a bus one day. A tremendous amount of wartime goods traffic moved after dark, in the blackout, so some friends and I would go 'spotting' by torchlight (whenever we could get batteries). Shining a torch along a 'Bulldog' one dark night we found it had no nameplate so we shone further along in search of the number plate and found instead an oval combined name-and-number plate, 3341 *Blasius*. Unexpectedly, we found streamlined LNER Pacific 4902 *Seagull* on goods one night, and later, a whole host of locos on foreign territory – Southern Railway 'King Arthurs'.

As the blitz on London eased off, so I cast my net wider and brought all the main line stations in the capital within my reach. My parents didn't object to my wanderings although they did warn me in a vague sort of way about certain dangers which lurked in London for a young lad yet, ever since I could remember, they had never curtailed my wanderings, even as a small boy. LMS 'Jubilees' and Midland Compounds at St Pancras, Southern 'Schools' at Charing Cross, the streamlined *East Anglian* and *City of London* at Liverpool Street – I met them all. Coming on to King's Cross station one morning, my schoolfriend Dave Tanner (who had now joined me in my wanderings) and I came face-to-face with streamlined 4489 'A4' *Dominion of Canada*, looking uncared for as it stood at the buffer stops. A friendly face appeared from on high and invited us up into the cab, where all the controls were pointed out as the engine stood there, simmering. With a short throaty blast on its melodious chime whistle, *Dominion of Canada* rolled the full length of the platform with me and Dave on board.

For one brief and glorious moment, all loyalties were with the LNER for, as friendly as Great Western drivers could be, they had never invited me on to that hallowed territory, the footplate – least of all, let me ride with them. This loyalty didn't last though, for soon I was back at the lineside at

Maidenhead with a tale to tell about a footplate ride on an 'A4' ('Oh yes, I "footplated" an "A4" the other day'), but it all paled into insignificance when compared with the fierce enthusiasm for the Great Western which was so apparent among the 'gang' I would meet there – all strangers to me but all with one common loyalty – the Great Western Railway. One boy, I recall, dared to refer to GWR engines as 'matchboxes on wheels'. Fists flew.

By now, I am sure, the 'steam bug' I had contracted had reached the incurable stage and the doodling started. Sketches of railway engines began appearing everywhere – in the margins of school textbooks, on the covers of exercise books, on scraps of paper – anywhere there was a blank space. Obviously, I enjoyed drawing but my art master didn't share my enthusiasm as I was considered by him as being 'only passable' in art. Often, as schoolboys do, we exchanged homework. Someone good at maths would take on the simultaneous equations set for homework and in return I would do his art prep. Invariably he would get a higher mark for his artwork than I would get for mine – yet I had drawn them both. One day, a much younger art master arrived to take over temporarily for a period which included exams, when a pair of boots were the test piece. Exams over, the results were analysed, and there was the pair of boots I had drawn being held up as an example of how a pair of boots *should* look. For the first (and only) time at school I had come out top in art. Quite simply, I wasn't one of the regular art master's 'blue-eyed boys' and I strongly suspect this was because of my working-class origins. We 'ordinary boys' were made to feel very aware of our place within the class structure of the grammar school I attended.

Geography was helped along by my interest in railways; history, too, as all the locomotives in the 'King' class (except 6028 *King George VI* and 6029 *King Edward VIII*) ran back in chronological order of the monarchs' reigns. All those castles whose names were born so proudly by Great Western engines must have a place somewhere either in England or Wales, and then there were all those coal wagons, stretching endlessly behind big 'Twenty-eight' class 2-8-0s, with names on their sides like Merthyr and Tirpentwys; they *must* come from somewhere. When we did the Welsh coalfields in geography I was a step ahead.

3283 *Comet*, one of the two 'Dukes' to retain their narrow cabs until the end of their days. The 3000-gallon tender is one of the early types as fitted to 'Dean Singles' etc, now with plates replacing the original coal rails and fitted with heavier springs (*Lens of Sutton*)

Not strictly a 'Duke' although still classified by the Great Western as being one, 3265 *Tre Pol and Pen* had been rebuilt with 'Bulldog' frames and was considered a very worthy 'cop' indeed (*Lens of Sutton*)

Back in earlier days, before the Great Western had really got me in its grip, I had been standing on Maidenhead station when a loco bearing a name unfamiliar to me, had come majestically steaming through – not fast, so that I was able to digest the whole of its long nameplate in one go. It was *Isambard Kingdom Brunel*. Who or what, I wondered, is Isambard Kingdom Brunel? My first thoughts, I remember, was that it was some Indian state, because of the 'Kingdom' bit. The answer came shortly afterwards when I stopped at the bookstall as I went to catch the train home, and found a copy of the Great Western publication *Track Topics*. By the time my train had drifted along the branch line into High Wycombe half an hour later I had a pretty good idea who this Isambard Kingdom Brunel was – Mr Great Western himself. In the rush to catch my bus to the village at Wycombe, I discovered to my dismay that I had left *Track Topics* on the train – a whole shilling down the drain, but from then onwards Brunel became my hero of heroes, and even to this day I am amazed at the sheer breadth of his thinking. This man, this little giant – not only did he tread virgin territory by planning and supervising the building of one of the new railways in the face of tremendous opposition, but he made his Great Western broader, flatter and faster-running than anything the world had envisaged at that time. What's more, he could go out and persuade people to put their money into his 'hare-brained' projects – hundreds of thousands of pounds. He had his failures – such as the South Devon atmospheric railway and, arguably, the *Great Eastern* steamship – but these were eclipsed by the enormity of his successes, in particular the Great Western Railway. Isambard Kingdom Brunel's name is proudly blazoned across one of my walls at home in a curve of polished brass and black, and my second son came within a whisker of being christened Isambard!

At home, books played a big part of everyday life. Both my parents were avid readers and I absorbed every bit of literature I could, especially if it was to do with railways. The Great Western was very generous with its publicity material and in time I managed to get hold of copies of all their 'shilling books for boys of all ages' (including a replacement *Track Topics*). The most popular of these was their *GWR Engines: Numbers, Names, Types & Classes* which listed all the Great Western named engines then in service.

Again I have my brother to thank, this time for introducing me to the magazine *Model Engineer*, copies of which he would bring home from the Technical School and which I would immediately grab and consume from cover to cover. One regular contributor to this magazine, and someone to whom I shall be forever grateful, was a writer who went under the pen-name 'LBSC'. He led me (and thousands of others) into the realms of the technicalities of steam locomotives and how they worked. Boilers, valve gears, injectors, and all the other 'blobs and gadgets' (as he called them) were explained away so that 'any Billy Muggins' (his words again) could understand their workings. What's more, his writings convinced me that I could build a miniature steam locomotive in the garden shed. Eventually he was proved right, though a good many years passed before I achieved this ambition and built a locomotive to the master's 'words and music'.

By now, the war was well under way, and during its darkest hours my father was keeping watch in the lookout hut which the Home Guard had erected atop the square churchtower, with his trusty old Purdy muzzle-loading shotgun across his knee. Not alone, of course, for he shared these lonely watches with other village stalwarts, among them Ernie Randall who, my dad told me, was a 'high-up' on the Great Western Railway. During these night-long vigils they must have discussed me and my passion for railways, for very soon I was invited to 'come and talk railways' with him. Ernie and I soon became firm friends as we chatted away in his little flint cottage at the other end of the village on long summer evenings. We talked the same language, Ernie and I – the Great Western Railway – and he held me spellbound as he recounted in detail how, at first hand, he had actually witnessed the sowing of the very first seed which had blossomed forth into the mighty 'King' class locomotives, by being present at the meeting of the Great Western's top brass when they had discussed the tractive supremacy gained by the Southern Railway with their new *Lord Nelson*, then took the decision to build a 'Super Castle' to compete, the locomotive which eventually rolled out of Swindon as the 'King'.

As we talked of 'Kings' and 'Castles', Ernie made it quite clear that, if I was interested, he could find me a job on the Great Western when I left school. Was I? I jumped at the chance of actually going to work amongst

3341 *Blasius* as station pilot at Exeter St David's in June 1949: the first of the straight-framed 'Bulldogs' and displaying the oval combined name-and-number plate on its cabside as picked out by torchlight in the wartime blackout by an enthusiastic band of young spotters (*W. H. G. Boot*)

4489 *Dominion of Canada* at the head of the 'Coronation' in 1937, taking on water at York and looking immaculate – a far cry from the grubby all-black engine on which two schoolboys 'footplated' a few years later at King's Cross during the Second World War (*C. C. B. Herbert*)

my beloved engines. Despite the war, you still had to have influence to get a job on the Great Western Railway, it appeared. Ernie's own influence for employment, years before, had come from no less a personage than the eminent lady doctor, Elizabeth Garrett Anderson, and now, steered by him, my goal was to have been an apprenticeship at Swindon, and although he waved the right magic wand and pulled off all the signals for me so that I became provisionally accepted, this was never to be. The first bombshell came when my parents learned that a premium of £100 a year was called for, and the next when they were told that I would have to be kept in digs. Then what about pocket money? The family budget just didn't run to this sort of thing. It was probably more than my dad was earning in those days.

Ernie, though, had other tricks up his sleeve, and although the result wouldn't involve me directly with locomotives, he did arrange for me to be interviewed by Mr A. S. (later Sir Allen) Quartermaine, the Chief Engineer (on the civil engineering side) at the Great Western's temporary wartime headquarters at Aldermaston. It was a fairly gruelling interview, I recall, with searching questions about my school work and spare-time activities. At Christmas, though, having achieved my matriculation (at the Great Western's insistence), I was immediately summoned by the Great Western for another interview, this time with Mr R. C. Kirkpatrick, the Divisional Engineer at Paddington, who had a vacancy for a drawing office junior.

A friendly uniformed man greeted me at the swing-doors at the end of the footbridge over Platform 1 that Saturday morning and ushered me into a hushed and leather-bound office in the bomb-scarred block alongside Paddington station. On the way we passed along a hastily repaired corridor where I almost tripped over a heap of old locomotive nameplates lying there. On the way back, the uniformed man and I looked at them more closely. *Iron Duke* was there, so was *Flying Dutchman* and a whole lot of old names out of the Great Western's glorious past. In the meantime, I had found 'RCK', a charming white-haired gentleman, considerably less formidable than 'ASQ' had been, more concerned about my welfare and whether I would be happy on his staff than my academic qualifications. Could I, he inquired kindly, manage to start work at Paddington on Monday morning? I certainly could – and did.

The Great Western Railway made no special effort to receive me with open arms when I joined it on that bleak, sleet-ridden Monday morning in January 1945. Instead, it treated me as being exactly what I was – a very new boy joining its vast and (to me) somewhat bewildering empire. Indeed, so little was it interested in me that the fact that I was late for work on my very first day had gone almost unnoticed, simply because nearly everyone was late that morning. Trains could be, and were, very erratic in wartime and my own (a LNER train) had come to a halt inside the tunnels leading to Marylebone station where it had stayed, hissing away to itself for a quarter of an hour, before limping into the station – a trick it was to play almost every morning, I learned.

Pens, pencils, erasers and the likes were issued to me, each item, to my secret satisfaction, indelibly marked GWR; even the duster I was given had the Great Western motif worked into its fabric. During the course of the morning I was handed a pink identity card which gave me access to anywhere on the Company's System (but not, it was stressed, on trains as a pass for free travel) and presented with a brand-new tissue-wrapped brass badge which I was to cherish and which showed the outline of what appeared to be a 'Castle' class locomotive and the words RAILWAY SERVICE: GWR. Very soon, staunch trade unionist Jack Rees had me cornered and was talking me into joining the Railway Clerks' Association. Jack, I noticed, was wearing a shiny silver-plated version of the same badge and I found myself wondering if such badges were reserved for trade unionists. Maybe that's why I joined.

My drawing board was the end one of a double row of some eight boards facing each other down the centre of a long narrow drawing office which lay somewhere within that large block of offices alongside Platform 1 at Paddington station. Each drawing board was a simple affair and consisted of a broad expanse of flat board covered in white cartridge paper which was propped at a shallow angle on the top of an old-fashioned mahogany desk. Another half-dozen or so similar boards were ranged along the further wall to catch what light they could from its windows, a number of which were still boarded over as hasty repairs to the damage they had received during bombing raids. From where I sat, though, I could just about glimpse

Relics of a railway-loving boyhood: elusive tinplate Hornby engines; the GWR books, 'for boys of all ages', *Track Topics* and *Cheltenham Flyer*; and three much-cherished and well-thumbed train-spotting books – *The ABC of Southern Locomotives* (*centre*) lost its cover years ago (*Brian Gough*)

Diesel railcar No 34, one of the Great Western's two express parcels cars, both of which plied in and out of the parcels platform at Paddington, seen here passing Old Oak Common East signal box in 1957. No 17, the one which I felt needed investigation at Paddington, had a more rounded, streamlined appearance but, unlike No 34, was not fitted with drawgear, being designed to work only as a solo vehicle (*R. C. Riley*)

through these windows and out over the top of Brunel's magnificent arched roof spanning the station, and where tarpaulins now flapped over holes made when Paddington station had received a direct hit from a bomb in March the previous year.

Shades of Brunel were everywhere. I had only to look just beyond my finger-tips to see evidence of him as a couple of inch-thick slices of his original broad gauge bridge rail acted as paperweights on my drawing board. Once, during my time on the GWR, I was privileged to be shown some old archive drawings, each one beautifully executed and looking more like a work of art than a working drawing. These, I was told, were the work of Brunel, and not only were they finely drawn but the lettering on them was exquisite. As the Great Western had started, it seemed, so it meant to go on, for my first task was not that of producing drawings but learning the art of lettering. Day after day, I sat there lettering away at my drawing board throughout those first weeks. A number of violent explosions shook the building as I sat there lettering – one, much too close for comfort, brought dust showering down from the lampshades on to the drawing boards and set the windows and teacups rattling. This was the tail-end of the V2 attack – enemy rockets which were still falling on London – and although one or two more distant rumblings were to punctuate my progress in lettering, the attack was soon to dwindle out entirely, much to my relief and that of all Londoners.

It wasn't all grind at the drawing board, producing alphabets, though. Being the office junior – and being made to feel very aware of it too – it was my job to run errands, most of which were in and around the station itself. These excursions I enjoyed as Paddington was the very hive of railway activity with always something or somewhere new to be explored, and one of my favourite haunts became the parcels platform extending beyond Platform 1 which offered a panoramic view over the whole of the lines leading to the station. I had gone there at first, I recall, to take a closer look at a diesel rail-car (No 17, a parcels car) which was being loaded and, as it was a type I had never seen before, I felt it warranted closer inspection. I poked my nose in everywhere but no one ever challenged me or asked to see my identity card, so I suppose my brass RAILWAY SERVICE badge sufficed. Altogether, I must have spent hours on the 'Lawn' (the passenger circulating area at the head of the platforms) which I could never cross without stopping to admire the magnificent model of the locomotive *King George V* which stood there in its glass case alongside the entrance to the Underground station. And always there seemed to be a full-size 'King' to be admired, standing at either the arrival or departure platforms.

The outer end of Platform 1 was always a good place to be, especially around 10.30 in the morning, as this was when the 'Limited' (the Cornish Riviera Express) went out. Up until the Second World War, the Cornish Riviera Express was known officially as the 'Cornish Riviera Limited' (hence the term, the 'Limited') due to its having a limited seating accommodation. The facility of booking seats was suspended during the Second World War and now, instead of running non-stop to Plymouth, the 'Limited's' first stop was Exeter, with the end two coaches being slipped at Westbury and Taunton. The Cornish Riviera Express and the Torbay Express (which ran on Saturdays only) were, I believe, the only two named trains running on the Great Western at that time. This was the Great Western's crack train – our pride and joy. Despite its wartime plain-green livery, there was no hiding the sheer physical splendour of the 'King' at the head of the 'Limited' as, seemingly effortlessly, yet at full stretch, it would dig its heels in, and with never a trace of slipping, start its heavy load of fourteen packed coaches and go buffeting away under Bishop's Road Bridge on the start of its long trek westward. I, and usually a knot of onlookers, would stand there happily until the last coach had drifted past, envying all those passengers their journey – even those standing in the corridors – and watch as the array of gleaming white discs and tail lamps of the end slip coach disappeared into the distance. Then it was about turn and a dash back to the drawing board for yet more alphabets.

Suddenly, one morning, things got better and I was eased away from lettering and onto drawing proper. Tracings at first, and these could cover anything from the siting of lamps on a station platform (gas as well as oil lighting was still widely used) to a new and complex track layout. The very first drawing I did for the Great Western was a tracing of a part of Southall loco yard, drawn to a scale of two chains to the inch and quite a challenge

Undisguised majesty: 6000 *King George V* displaying its transatlantic brass bell and cabside medallions, though looking a trifle neglected, as it stands at Exeter, still in wartime plain green livery in September 1947, some three months before nationalisation (*R. C. Riley*)

Perhaps Virol would have helped on the morning I crossed that seemingly formidable maze of tracks. 5042 *Winchester Castle* passing alongside the vast goods station as it approaches the platforms at Paddington with the Cheltenham Spa Express in June 1951, some ten years after it had become my first serious 'cop' (*Brian Morrison*)

(to me, anyway) as it contained a fair amount of switches (points) and crossings. I must have made a reasonably good job of it for I was then set to work on another tracing, this time a general arrangement of a Pooley weighbridge for installation in station yards for weighing loads as they went out on the backs of lorries or horse-drawn carts. Prints were made on white paper from these tracings, then a watercolour wash applied to indicate the various applications or materials shown on it. All new work to be carried out was coloured red (carmine), I recall, and a right old mess I made of those early attempts until I got the hang of applying a colour wash by using plenty of water.

It was Reggie Collins who first managed to slide me over on to drawing and, due to pressure on his work load, I was soon to become his assistant and he my mentor – and friend. Officially, though, I was there to do tracing and drawing for anyone in the drawing office who needed it. Reggie was responsible for a whole host of the minor facets of civil engineering which came under the umbrella of the Divisional Engineer's Office – such things as drainage, the routing of water mains under the track, telegraph wires over the track, and even such things as the siting of advertisement hoardings – they were all Reggie's responsibility. Like so many at that time, he was outwardly discontent with his lot, a combination of the strain of wartime overwork and low pay. The Great Western Railway, in common with other railway companies, was not over-generous with its salaries, particularly, it seemed, to the lower echelons of its professional staff, my own pay at the age of fifteen being 30 shillings a week. Nevertheless, and despite his discontentment, Reggie had a loyalty to the Great Western Railway – an underlying pride – and as soon as he became aware of my own deep interest in all things Great Western he went out of his way to foster it.

Everyone knew Reggie, from those in high office to the platform staff – and not only at Paddington for, wherever we went together on the System, everyone had a cheery word for him. And he had a maxim: 'Always ask,' he would say. Anything was possible if you *asked*, though it hadn't helped him much with his pay rise. Nor did his maxim always work, as I recall applying it one day when I asked the driver of 'Castle' class loco *Earl of Plymouth* if there was any special reason why his engine should be running with the unique eight-wheeled tender coupled to it.

'Well,' replied a soft Welsh voice, 'we've got to keep our bloody coal and water somewhere, 'aven't we, boyo?'

So much for Reggie's advice, I thought.

My career took another step forward one morning in early March when Reggie decided that it was high time I came with him on an outside job – not very far away – on the station, in fact, where I was to help him take one or two measurements. Armed with a leather-cased 100-ft tape-measure and my brand-new, and so far unused, bright-yellow-covered surveyor's notebook, we sallied forth and made our way along Platform 1. Policemen, porters all had time of day for Reggie, and a girl in the bookstall waved as we went by. Reaching the end of the platform, he led me up to the signal box where, in response to his knock, a friendly face greeted us and we were invited inside. There, Reggie stood aside as he handed me over to one of the signalmen who worked away at his array of switches and coloured lights (Paddington's signals had, by then, been all-electric for some years) as he explained the intricacies to me, showing me how he was giving the 'green' to the 'Limited' as it was about to depart. From this new and fascinating vantage point I watched the 'Limited' heave its way out of the station and pass by, right outside the window, then follow its progress on the illuminated overhead panel as it went on into the next section down the line.

'Lesson one,' said Reggie as we left the box. 'Always ask, and you can go almost anywhere – even into signal boxes. Now for lesson two.'

To anyone who has ever stood looking outwards at the far end of Paddington station, the maze of tracks must look formidable. It did to me that morning as we jumped down from the platform and started to cross the line. As Reggie pointed out, from now on I was going to spend a lot of time out on the line, so I had better get used to the idea and he was throwing me in at the deep end. Under his guidance I crossed to the far side, taking one track at a time, keeping one eye on traffic movement and the other on the state of the signals before moving on to the next track, 'freezing' in the 'six-foot' (the clear area between running rails) if anything should come

The subject of the backing-sheet doodle: 2935 *Caynham Court*, one of the 'Saint' class which was fitted experimentally with rotary-cam poppet-valve gear, seen running light through Didcot en route for Swindon in 1934 (*Dr Ian C. Allen*)

An old stamping-ground where I trespassed, unchallenged, many a time: the interior of Old Oak Common roundhouse with (*l-r*) 4080 *Powderham Castle*, 6937 *Conyngham Hall*, 6947 *Helmingham Hall* and 'Forty-seven' class 2-8-0 4703, April 1964 (*R. C. Riley*)

along. It did. A pannier tank came fussing out of the station, shrieking its whistle at us. The driver made a gesture. Reggie waved back. He knew everyone.

Regaining the safety of Platform 8, well pleased with our progress, we strolled along the cab road then back round to Platform 1 where we went into the tea room for a cup of coffee, OCS (On Company's Service, which meant a penny off). While we were taking our ease, Reggie chatted to the girl behind the counter. She knew him, of course.

Quite a morning, we agreed, as we headed back to the drawing office. Then we realised that we had completely overlooked the mission we had set out to accomplish, so it was back to the station and along to the end of the middle platforms. It was a straightforward job and didn't take very long – the re-positioning of a sign warning passengers not to cross the lines.

Spring came, and suddenly everything happened at once. R. C. Kirkpatrick retired and E. T. Davies took over the reins as Divisional Engineer at Paddington. The war in Europe ended, the Chief Engineer's Office decided to come back to London from Aldermaston and took over our office space so that we had to move out, and I 'went solo' on my first outside job. This was at Royal Oak, the first station down the line from Paddington, where I had to meet the local permanent-way inspector and take measurements for the siting of some new fencing.

'Just a ten-minute job,' said Reggie, but Ranelagh depot, where engines were rostered and turned for backing onto trains at Paddington, was close at hand so I had a good look round and it was two or three hours later before I walked back along the line, crossed the parcels depot, and made my way up the ramp where the big carthorses plied and across the street to where our newly acquired offices stood. This was in Orsett Terrace, some five minutes' walk from the station, in a compact three-storey building next to the church where we, the drawing office, occupied the three rooms on the top floor. I shared one of these rooms with Reggie, Fred Cockhill and a small elderly man called George Fear, who looked after all the plans and did our typing for us. It was a cosy arrangement after the larger and impersonal office we had all shared, back there on the station.

A door led out on to the fire escape where there was a platform surrounded by an iron railing, and from where Reggie would hold long conversations with the vicar's wife next door, or regale passers-by in the street. At other times, he was a ship's captain pacing the bridge, addressing the world at large with 'Sink me that ship, master gunner'. Calling me one day, he said, 'Look at that – he's allowed to do that you know.' One of the carters from the parcels depot opposite was relieving himself in the street against the offside wheel of his cart while his horse patiently waited. Reggie hailed him. Without stopping what he was doing, the man turned and waved back. Everyone knew Reggie.

Partly because of this new-found freedom, and partly because I was in such close contact with locomotives, the doodling started again. The broad expanse of clean white cartridge paper on my board had been just too much for me and, soon, little snippets of drawings – details of locomotives I had seen – began to appear along the bottom edge, then creep up and over the broad surface. 'Saint' class 2935 *Caynham Court* had been really to blame, as I had seen it down on station where I had inspected it at closer quarters and decided that its one-off Poppet valve gear needed further explanation – on my backing sheet, along with a view of 3448 *Kingfisher* which I had seen come in on a double-header. It was good to see a 'Bulldog' at Paddington – we didn't see many of them.

In overall charge of the three drawing offices on the top floor was Mr Hudson ('Huddy'), a very tall, pipe-smoking, quiet and (usually) gentle man. One particular lunch hour – a Friday – I had dashed across to one of my favourite shops which went under the name of 'Bonds o' Euston Road', and which was a positive Aladdin's cave where every conceivable item to do with model steam engines and railways was to be found on its packed shelves. Getting back some ten minutes late, I had just opened the long package I had bought at Bonds when Mr Hudson came storming in and demanded to know where I had been and wasn't it high time I put down a clean backing sheet. I had never known Huddy so cross. Then he caught sight of my half-open package – rails, chairs, sleepers and all the makings of half a dozen yards of gauge 'O' track I had hoped to put together over the weekend. Very soon, Huddy had the rest of the permanent-way section in

'Right away' for a 'Marlow Donkey' at Marlow station, June 1952. A selection of 0-4-2 tank engines of the 48xx (later 14xx) class worked this Thameside branch from Bourne End, each one youthfully referred to as a 'Donk'. 1448 was one of those fitted with boiler top-feed, directly behind its chimney (*R. C. Riley*)

One of the handsome little Lambourn branch engines, 1335, at Reading shed, June 1937. Three of these 2-4-0s were taken over from the M&SWJR by the Great Western who re-boilered them but kept their distinctive cabs and original tenders (R. C. *Riley*)

from the office next door and he, and the rest of them, were laying gauge 'O' tracks across my board. The following Monday, Huddy handed me a book, a 1920s edition of the Bassett-Lowke *Handbook of Model Railways*, which he had bought when he started on the Great Western all those years before. We became great friends, and although I did my best to maintain a clean backing sheet, I never quite got out of the habit of doodling.

There were a number of model railway shops in London in those days, each worthy of a visit, I considered, and each with a charm of its own which was almost old-world. There was Mills Bros. in Southampton Row opposite where the trams dipped underground and went clattering away under the streets, and Bassett-Lowke's in Holborn, with its junction signal bracket over the door, set to indicate a clear road into the shop. Both of these shops were convenient to call on when I had to take tracings for printing at Colyer & Southey's in Tooks Court, just off the Holborn end of Chancery Lane. Nearest to hand, though, was Walker & Holtzapffel's in Baker Street, where a whole range of hand-made model locos were always on display, watched over by an elderly gentleman who knew I had little money to spend but who never objected to my gazing longingly at the contents of the showcase at the far end of the softly carpeted shop.

It was my interest in these fine locomotives which landed me in a spot of bother one day. Reggie handed me a sketch which had been sent in to the department and which showed the siting of a 'To the Station' sign which, it was suggested, should be attached to a lamp post in the middle of the road at the junction of Praed Street and Eastbourne Terrace, outside the Great Western Royal Hotel. Reggie asked me to step along, just to check that it would be all right to put the metal sign in the position shown on the sketch, so I went. A cursory glance at the lamp post, and I thought it seemed just the place for a sign, then made the most of my freedom and sneaked away for half an hour's browsing at Walker & Holtzapffel's glass case display before returning to the office. A few days later, a mangled-up 'To the Station' sign landed on Reggie's desk. A double-decker bus had taken the turning from Eastbourne Terrace into Praed Street rather wide and the sign had gone flapping all the way along the length of its upstairs windows. Disdain was all that Reggie could find for me that day, but it didn't last.

Improved lighting was a high priority now that the war had finished, and one morning I went to help Reggie in the siting of some new lamps in the loco yard at Old Oak Common. These were floodlights, 'parabolics', mounted high on tall poles. To get there we had to catch the 'empties' (empty carriages) as they went back from Paddington to Old Oak, a distance of some three miles. Having asked the driver ('Always ask . . .') of the engine if I might ride with him on his footplate, Reggie settled back in a first class compartment while I travelled tender-first on a 'Hall' class engine, to arrive at the other end, rather grubby but very pleased with myself at having 'footplated' a Great Western engine at last. I was to go down to Old Oak many a time after that, and often I would ask the driver if I might come up with him. Some said no – it was their prerogative – but on the whole I found enginemen a friendly bunch and seldom 'rode the cushions' down to Old Oak Common. Once, I spent a whole Saturday afternoon there, browsing round the loco shed and watching engines being repaired, and was more than willing to give a hand if required – but no one asked – least of all to see my identity card.

Soon, though, my work was taking me much further afield, on branch lines as well as on the main line. The London Division embraced a whole host of cross-country lines as well as branch lines, most, alas, now gone for ever. The majority of branch lines were worked by those handsome little 48xx (later 14xx) class 0-4-2 tank engines (still, to me, 'donks', as one of them worked the Marlow branch near my home and was always known locally as the 'Marlow Donkey'; the 'Donk' to us youngsters), but on the Lambourn branch I met the handsome little 2-4-0 tender engines with their origins in the Midland & South Western Junction Railway, while on the DN&S (the Didcot, Newbury & Southampton line), 'Dukes' and 'Bulldogs' still roamed freely. Still my favourites, these, though I must admit to having a soft spot for the little 2-4-0 'Metro' tank engine, a long way now from the Metropolitan Railway in London from whence its class ran originally, as it pottered along the edge of the Cotswold countryside on the Fairford branch.

These branch lines were lovely places to visit on fine days in summer, less so in winter, and whenever I had to go anywhere along the Fairford branch

A long way from the Metropolis – and far removed from the Fairford branch whose Metro-hauled train I delayed: a Metro tank, 3581, at Taunton, September 1936 (R. C. Riley)

One of them missed me by inches: an unnamed wartime-built 'Modified Hall' 6965 (later named *Thirlestaine Hall*) in plain all-over black livery and without cabside windows, at Newport in 1946. Gas lamps, such as that on the left, abounded on the GWR and were included in the miscellany which came under our department (*R. C. Riley*)

it always seemed to be winter time – and raining. The very first time I went there – on a wet morning – I missed the train from High Wycombe to Oxford (for the Fairford branch) so I carried on up to Paddington then took a train to Swindon where I hoped to catch a bus out to Fairford. I caught one all right – it went to Cricklade first, then made inroads into rain-soaked Gloucestershire to Cirencester before heading back towards Fairford. Eventually it got me there, at two o'clock, just as the afternoon train, a 'Metro' and a couple of ancient coaches, was about to leave.

'Hold the train!' bawled the permanent-way inspector as he saw me come splashing across the yard and on to the platform. He wasn't very happy – I could tell straight away. Not only had he waited around all the morning for me, but now he had to paddle about in the long grass behind the station in the rain as we located a septic tank and took measurements of its underground pipe from the station, due for renewal. What's more, he insisted, it had all been measured up before. The 'Metro' waited patiently, its steam beating down across the platform in the rain. Half-an-hour later, it was allowed to leave, bearing me, wet through and clutching my soggy notebook, sitting opposite an equally soggy permanent way inspector, who peered out of the misty window and spoke not a word on the hour-long journey back to Oxford – while the gallant little 'Metro' galloped along between stations in an effort to make up for the lost time I had caused.

Back in the office next morning, dried-out and ready to convert the previous day's measurements into a working drawing, I turned up the file on Fairford station. No sign of the inspector's measurements there. They did turn up though – weeks later, in Reggie's drawer. Give or take the odd half-inch, they tallied up with my measurements.

Working for the Great Western wasn't all railways though. Company-owned roads and station yards had to be attended to as well as things like stables – and then there was the Kennet & Avon Canal. This more or less followed the main line (the Berks & Hants line) from Reading and out of the London Division's territory somewhere down beyond Hungerford. It was no longer in use commercially, but still the Great Western's responsibility, and regular reports were sent in by the permanent-way inspectors through whose territory it passed, indicating the state of the weeds and degree of silting up, and which warranted a visit from a member of our staff periodically, just to lend credence it seemed. There were always keen anglers ready to volunteer for this duty but when my turn came round I decided I would like to go. I knew nothing about canals – nor, I suspect did the permanent-way inspector I met that morning, somewhere down near Kintbury, so we walked along the towpath in the summer sunshine, peering at the weeds and disturbing the clear water with sticks. At lunch time the inspector steered me towards a pub where, on the counter, a penny-operated Polyphon (a device like a large musical box) played but one tune, 'The Honeysuckle and the Bee', over and over again, much to the annoyance of the locals. As we left, we put yet another penny in the machine and went back along the towpath, well refreshed and with the strains of 'The Honeysuckle and the Bee' floating on the balmy afternoon air. No mention of this musical interlude was made in my report – simply that I agreed with the inspector's findings and that the canal at that point was not navigable.

There was, however, a sombre side to our duties at Paddington, and one that we all found disturbing – that of preparing accident reports – more disturbing to me as I was often the one to answer the telephone call notifying us of the incident. It was Fred Cockhill's job to make site plans of all accidents on the London Division, fatal or otherwise, and I could always tell when he was off to a coroner's inquest as he would come to work in his dark suit and long down-to-earth navy-blue overcoat. Accidents on the line were not infrequent, and I recall the shock one day when we learned that one of our own permanent-way inspectors – a life-long railwayman – had been run down and killed by a train. Another time a child had wandered on to the foot-crossing at Old Oak Lane Halt on a Saturday morning and had been struck by the 9.10 express to Birkenhead (I took the phone call for that one), and then there was the incident of the ganger whose severed head was found under the buffer beam of a locomotive as it had pulled into a station.

I almost became the subject of an accident report one day. It was one of those bleak, raw cold mornings in December, just before Christmas, when a thin mist hangs everywhere, and I had gone with Vic Hobbs of our permanent-way section to measure up a set of switches due for renewal on

The 'Whitewash': the track-testing car, an old GWR 'toplight' brake third coach of 1911, specially adapted in 1932 as a simple means of detecting track irregularities by dropping splashes of whitewash on the line. Surely now the oldest carriage running on BR, it was re-bogied in 1980 for 100 mph running (*R. C. Riley*)

The 'Limited' through Reading: the Cornish Riviera westward bound on 12 January 1953, headed by 6009 *King Charles II* and taking Reading General station in its easy stride, shortly before swinging off along the Berks & Hants line (*R. C. Riley*)

the fast main line west of Slough station. Because it was such a busy section of track, we had to have a lookout man with us so we had picked a local man at Slough station, and he came along the line to safeguard our interests while we concentrated on taking our measurements. To get the accurate measurement needed for the blade (the section of the rail that moves to switch trains) I had, without thinking, put my hand in the gap between the blade and the rail where I was holding the end of the tape, and where any movement of the blade could have trapped my hand or even crushed it. Noticing my folly, the lookout man turned and gave me a full broadside of railwaymen's language, and Vic, way down at the other end of the tape, looked round to see what all the fuss was about, then joined in with his own choice of invectives. In my confusion, I happened to glance up – just in time to see a train bearing down on us, coming full-pelt out of the mist. I shouted – we all jumped clear – and a split second later the train went shrieking past. To my dying day, I shall remember that locomotive – it was one of the unnamed 'Modified Halls'.

We had lunch in the gangers' canteen near the station – sheeps' brains and barley, I recall – and as Vic and I sat there, the lookout man joined us. He grinned. 'We both learned a lesson this morning, didn't we, son?'

The Great Western was teaching me something new every day.

Later, I had another 'near miss' with Vic, and one that I shall always regret. He came bursting up to my board one morning and thrust a memo under my nose – one, which he said, would interest me. The Great Western was, at that time, investigating the possibilities of employing four-aspect signalling, a system which was already in use on parts of other railway companies, and which gave an indication of the 'clearness' (or otherwise) of the line much further ahead of a train than the two-aspect signalling being employed by the Great Western, thus suiting itself to train-running at much higher speeds. This new system meant a modification to the ATC (Automatic Train Control) with which Great Western locomotives were fitted and which was operated by an electrically energised ramp fixed between the running rails.

According to the memo, a high-speed train was to be run between

Reading and Maidenhead to test the application of the ATC in conjunction with four-aspect signalling, to be headed by a 'Castle' class locomotive in tip-top condition and its tender full of best Welsh steam coal. Vic, as a member of the permanent-way staff, was to join the trip, and saw no reason why I shouldn't go along as his assistant – unofficially – just for the ride.

The day chosen for the run was a Sunday, when traffic would have eased off and the main line comparatively clear. I was to pick up the outward-bound train at around 9 a.m. but the problem for me was that of getting to Maidenhead for the time of departure as no buses left the village until midday on a Sunday, so my bike was decided upon as a means of getting me there at the appointed hour. This was a normally trusty machine – a BSA, fairly lightweight and maintained by me in reasonably good condition. I had a shrewd idea of the cycling time to Maidenhead as my bike and I had made the trip together many a time during my train-spotting days, and being Sunday, there would be little or no road traffic about. Maybe I cut things a little too fine, but I seemed to be well on schedule as I paid my dues and crossed the toll bridge at Cookham. Head down, and coming past Bolter's Lock at speed though, disaster struck: a puncture. I prided myself that I could whip a tyre off and mend a puncture at the roadside in record time, but not that one – it eluded me completely. Cursing wartime tyres, I put the cover back on and pumped up. By the time I had reached the Bath Road and turned into Maidenhead, it was flat again. By a series of pump-ups, I covered the last mile to the station – just in time to hear the staccato bark of a 'Castle' starting off, intent later on putting up the performance of its lifetime.

I must say, Vic was very good about it next day at work, although he did rub salt into my wounded pride when he told me they had topped the 'ninety' mark on the run. And he did his best to make up for it as, later on, he fixed it for me to travel with him on the 'Whitewash', down as far as Westbury. The 'Whitewash' was a specially adapted coach which was attached to the rear of a scheduled train, and which located any irregularities in the track, to drop splashes of whitewash on the sleepers as it passed over them in order that they could be easily located by the permanent-way gangs. It was an interesting and enjoyable day out, but

Looking every bit 'at home' on a passenger train, 4704, one of the big 'Forty-seven' class 2-8-0s (originally intended for fast vacuum-fitted freight duties) handles a Saturday express, the 1.25 p.m. Paddington to Kingswear, down Bruton bank, on 18 July 1959 (*R. C. Riley*)

Outward bound from Paddington, semi-fast to Maidenhead then round the Wycombe branch: 'Sixty-one' class 6141 in August 1953 and little changed from GWR days except for BR emblems, front number plate and grabrails added to boiler strap between chimney and safety-valve (*R. C. Riley*)

nothing could ever make up for missing that high-speed run.

Naturally, I came in for a lot of ribbing over my cycling misadventure, so to save face I decided to cycle the 30-odd miles to Paddington one Saturday morning. This time, all went off without a hitch, and having negotiated the remains of the tramlines still left here and there in the road between Uxbridge and Shepherd's Bush, I pedalled up at Orsett Terrace unheralded. This time I was half an hour early.

Everyone at Orsett Terrace, even the girls in the clerical department downstairs, must have been aware of my keen interest in trains, especially the high regard which I had for the Cornish Riviera. Seeing it depart remained a highlight of my day throughout the whole of my stay with the Great Western — that, and watching the up 'Limited' roll to a standstill under the wide-arched roof every afternoon — invariably on time. There was a saying all the way down the Great Western Railway that you could set your clock by the 'Limited', and I believe it to be true. As often as possible, I would sneak away from my drawing board at 10.29 and pop across the road to the parcels platform, just to watch the 'Limited' go out. It took only five minutes but so often during these clandestine breaks the Divisional Engineer would ring for me to go down to his office (we had a system of bell codes — mine was four rings, pause, two rings). Knowing I wouldn't be there, he would regard whoever had gone down to make my excuses over his half-moon specs, get out his watch and say, 'Ah! He's watching the "Limited" go by.' This became a kind of catch-phrase, and whatever the reason for my absence from my drawing board (at any time of the day), when someone asked, 'Where's young Wheeler?' the answer would come back, 'He's watching the "Limited" go by.'

I still had this yearning to travel on the 'Limited', and it was John Newton who decided to do something about it. John was a colleague of Vic in the permanent-way department, and he had a hankering, he said, to go back to St Ives where he had spent holidays as a lad in pre-war days, so it was agreed that he and I should take a long weekend down there, Friday till Tuesday, at Whitsun. Now, after twelve month's service, I was entitled to four free travel passes a year — and what better way to use my first one than on the Cornish Riviera.

On a beautiful morning in early summer I realised my ambition, and John and I boarded the 'Limited' – and away we went to the West Country in style. By now the 'Limited' was restored almost to its pre-war splendour – not quite though, as the extra-wide Centenary Stock coaches had not yet been put back on the train (they still turned up singly on all sorts of trains), but it was now equipped with refurbished modern stock throughout, complete with restaurant cars. Up front was 6021 *King Richard II*.

The heavy 'Limited' was never one of those trains which galloped into its stride – rather, it strolled into a pace then settled down at a steady 60 mph or so, with dignity. By Slough we were bowling along well into the 'sixties' with no fuss at all, easing off a bit at Reading where we swung left and struck out westwards along the Berks and Hants line, with the Kennet & Avon Canal keeping us company alongside. Despite our speed, it all seemed so leisurely. Every now and again I checked the pace with my wristwatch (specially bought with what little savings I had, for times like these – it had a sweep second hand, quite a new thing in those days) just to make sure we were travelling as fast as we were supposed to. At Westbury, John and I hung our heads out of the window and watched the first slip coach come off, then the second at Taunton where I caught a glimpse of 'Bulldog' 3453 *Seagull* standing in the bay platform. Beyond Taunton we climbed up to Whiteball summit and as we did so John pointed out the Wellington monument (he was a mine of information) so I retaliated by pointing out that this was where *City of Truro* had made history. We lunched Great Western style on fish (it was Friday) on green newly upholstered swivel chairs (third class), took Exeter through the middle road non-stop and stood in the corridor at open windows waving to the promenaders as the 'Limited' hurried along the seawall through Dawlish Warren. A stop not in the timetable brought another 'Bulldog' to our aid at Newton Abbot and together John and I went up to the front of the train just to hear the 'music' as 'Bulldog' and 'King', coupled together, took us up and over Dainton Summit, at little more than walking pace. At Plymouth we stretched our legs on the platform to see our two locos come off and a 'Castle' take their place to haul us the rest of the journey through almost the length of Cornwall, one which started with the highlight of the

1000 *County of Middlesex* with its original double chimney and flush-sided tender but now in BR green livery (cabside and cylinder lining noticeably different from Great Western lining) at Old Oak Common shed, May 1956 (*R. C. Riley*)

LNER V2 class 4844 *Coldstreamer*, one of the handful of named V2s, in its original pre-war apple-green livery. They were always in an all-over grimy black when I travelled behind them but they certainly ran well (*C. C. B. Herbert*)

whole trip – the crossing of Brunel's magnificent Royal Albert bridge at Saltash. There never was such a journey as this one.

Slender, almost fairy-tale-like viaducts led us over steep valleys, stations sprouting palm trees on their platforms went drifting past and china-clay tips with the ghosts of ancient engine houses interrupted my new horizons from the carriage window as we travelled through Cornwall on this far-western thread of the Great Western. Truro Cathedral, majestic and mellow in the afternoon sun, hove into sight – and soon afterwards we eased into St Erth station where we exchanged the luxury of the 'Limited' for an impudent local train headed by a 'Forty-five' tank which skirted us around the peninsula on the edge of a blueness such as I had never seen before, and into St Ives.

I recall little of the rest of that short holiday, save that of gazing out over the sea to St Michael's Mount and, on the Whit Monday Bank Holiday, standing soaked to the skin, in a steady all-day Cornish downpour on the edge of the Loe Pool near Helston – and Mrs Eddy, our homely Cornish landlady, whose address we had got out of a remote cycling handbook, and who fed us well on fresh mackerel. I don't even remember very much about the journey back to London on the Tuesday, although we travelled on the up 'Limited' – that I know for sure.

Next day I watched the 'Limited' go by with a new sense of pride – I had actually *travelled* on it.

If the impression given so far is that my time on the Great Western was one big out-and-about adventure watching trains, nothing could be further from the truth. By far the greater part of my time was spent at the drawing board working seriously, particularly now that the war was behind us and there was a general feeling abroad on the Great Western of 'Let's get back to normal'.

In the Divisional Engineer's Office we became involved in such things as new stations – one was projected and drawn up (then abandoned) at Ruscombe on that long stretch of main line between Maidenhead and Twyford, while in the permanent-way department more and more sections of track were being relaid with the new and heavier flat-bottom rail to replace the more traditional rail of bullhead section. On the running side, things were moving ahead, too, as locomotives started appearing in their old pre-war lined-out liveries, new coaches were being built and older ones smarted up with a new coat of chocolate-and-cream paint instead of plain brown, and handsome twin-shield Great Western crests now adorned their waistlines. Refurbished restaurant cars with fluorescent lighting and etched perspex panels between dining tables began to make their appearances on more and more trains.

Travelling became easier, though I have to admit to having seen Paddington station on Saturday mornings at the height of summer completely clogged with travellers who formed queues three or four deep which stretched for over a mile, out of the station and around the adjoining streets. War-weary folk were determined to take a well-earned break, but by early afternoon, and by running scores of extra trains, the Great Western got them all away safely. Here, the big 'Forty-seven' 2-8-0s, usually busy on fast freight, gave a helping hand on passenger trains, I recall.

I came across another incident of one of the Great Western's big 2-8-0s on passenger work, only this time it was one of the 'Twenty-eights', engines which normally handled the extremely ponderous and somewhat slow mineral trains. By now I had forsaken the LNER train with its tantrums in the tunnels leading to Marylebone and was catching an earlier train from High Wycombe – a Great Western – which was getting me to Paddington half an hour early and thus allowing me a leisurely saunter through the station and along the parcels platform before turning out into Orsett Terrace to start work. Invariably, this train was worked by one of the 61xx class tank engines ('tanner-oners', as we called them as boys) and if ever there was an engine which not only looked the part but played its role to perfection, it must surely be the 'Sixty-ones'. They were the ideal suburban passenger train engine – rugged and handsome, with a turn of speed coupled with an acceleration which was hard to match – and more than equal to the getaway of the electric Underground trains which plied in and out of the far side of Paddington station.

Many's the time I would travel home to High Wycombe the long way round on a Saturday afternoon, via Maidenhead and the Wycombe branch,

A 'shufflebottom'. The Southern Railway's post-war image in the form of one of its 1945-built 'West Country' class locomotives. 21C113 (later named *Okehampton*) at Exeter Central, September 1946 (*R. C. Riley*)

Single-chimneyed 1012 *County of Denbigh* in BR black livery at Plymouth Laira shed, September 1954. 'A bit unsteady on her feet,' the driver had said when she was brand new and without a name, simply plain 1012 (*R. C. Riley*)

just to experience the performance of a 'Sixty-one' tank out on the main line. Running fast to Slough, it would top the 'seventy' mark – easily – as it galloped away from Paddington and the driver gave it its head on the open road. After Maidenhead it would settle down and enjoy an easy gentleness round the branch line, seemingly every bit at home in these rural surroundings as it stopped at all the stations, before charging with a final spirited burst up the steep incline and come to rest in the bay platform at High Wycombe.

One morning, though, our 'Sixty-one' was definitely not in good spirits – in fact it limped into the station and heads were scratched as to what was to be done with the train with the ailing engine. Then it was decided to hand over the train to the big 'Twenty-eight' which was standing on the middle road with a long freight train which stretched its way throughout the length of the station and out of sight under the road bridge. The 'Sixty-one' came off, the 'Twenty-eight' was coupled up – and we were away like the wind. It was one of the best runs I have ever experienced to Paddington from High Wycombe, with the small wheels of the big 'Twenty-eight' buzzing round like clockwork between stations, doing its best to make up for lost time, and although it tried hard to get us into Paddington on time, it didn't quite succeed. Anyone standing on the platform at the terminus that morning, though, would have witnessed a rare and spectacular sight – a big 'Twenty-eight' freight engine on a suburban train, gently rolling to rest at the arrival platform – almost on time.

Quite suddenly, and completely out of the blue, a new locomotive appeared at Paddington. As I strolled up the parcels platform one morning I saw in the distance an engine coming tender-first towards the station from Ranelagh yard. That tender – it was different – plain, with no familiar flair along its top edge. My first reaction was that it must be a 'foreigner', a loco from another company, but as it drew nearer there was no doubting it was a Great Western engine – a new one, with the number 1000 on its cabside and looking a bit like one of our 'Modified Halls' except that it had a single long splasher over its wheels and – of all things – a copper-capped *double* chimney. As it drew level, the safety valve of No 1000 lifted with a tremendous roar which sent the pigeons on the platform clattering away into the roof. Retracing my steps to the station I went along the platform for a closer look at this newcomer as it backed on to its train. A glance up into the cab showed the pressure-gauge needle hovering around the 280 psi mark instead of the usual 250 for a 'King', or 225 for a 'Castle'. So – we had gone 'high pressure'. No wonder those safety valves had gone off with such an ear-splitting roar.

Rumours had been filtering through from Swindon for some time now about a new locomotive 'on the stocks' so I wasn't altogether taken unawares by the new engine. What did surprise me, though, was that it was another 4-6-0, as we had been led to believe that the new one would be a Pacific – a 4-6-2. This, I was told, would be a comparatively lightweight engine with a wide firebox to make better use of the inferior coal we were getting at that time, and an axle-loading which would allow it to run through all the way from Paddington to Penzance. The class, it was rumoured, was to have been the 'Cathedrals' and now, instead, Swindon had sent us these new 4-6-os – the 'Counties' as they later became. Over the next few months, more of these new engines appeared and they were to become a familiar sight at Paddington, though the first one, No 1000, was the only one to have a double chimney – in fact, she was the only locomotive on the Great Western ever to receive one, though others, including the 'Counties' were to get them in BR days.

Now that the railways were getting back into gear again I decided I ought to take a look 'over the fence', just to see how the other companies were getting on. As a railway employee I was allowed privilege tickets at a quarter of the normal rate to travel on all the railways, so one Saturday afternoon I 'trespassed' on the LNER and rode from King's Cross to Peterborough behind an 'A1' Pacific, then came back behind a 'V2' – a 'Green Arrow' – at, I must say, a very impressive rate, as I timed it all the way. I had for a long time been an admirer of Sir Nigel Gresley and his engines, ever since I learned how he swallowed his pride and admitted his defeat in the GWR/LNER locomotive exchanges of 1925, then applied the lesson learned by adjusting the valve gear 'Great Western fashion' on his 'A1' Pacifics – to produce a truly accountable and free-running engine. This historical exchange resulted in LNER No 4474 coming to the Great

Besides the 'Castles' mentioned (5039 *Rhuddlan Castle* and 5091 *Cleeve Abbey*) three more were converted to oil-burners. 100 A1 *Lloyd's*, seen here at speed on an up Weymouth express near Maidenhead in 1947, carried its oil supply on a high-sided 4000-gallon pattern tender. The tank top is visible, so too is the sliding shutter over the cab cut-out, a feature of the oil-burners (*M. W. Earley/NRM*)

Post-war inferior coal in evidence: wartime-built 'Hall' 6933, as yet unnamed and still without cabside windows (not fitted as a wartime blackout expedient) smoking well at Reading, July 1946. It was given the name *Birtles Hall* the following December (*R. C. Riley*)

Western and 4079 *Pendennis Castle* taking to the LNER rails. Whereas the performance of 4474 was far from being considered a failure on Great Western metals, *Pendennis Castle* excelled itself and caused many a raised eyebrow because of its coal economy and brisk performance. In all fairness, though, plans were already afoot for modifying the valve gears of the LNER engines before the exchange took place, but it can be said that it was the performance of the Great Western engine which prompted the LNER to take action. And was not one of them, 4472 *Flying Scotsman*, to become the first British locomotive to be *officially* timed at 100 mph? His 'V2s' were extremely versatile as well as being, I think, very good-looking engines, and these were often employed on the Saturday 12.15 Marylebone to High Wycombe 'fast'. A mad dash to Marylebone station, usually made by jumping on a No 27A bus as it passed the end of Orsett Terrace and I could just about make the 12.15 'fast' by the skin of my teeth.

By this time, I had travelling companions – young people of my own age who also went home by the same train, and it was the duty of the first one on board to grab the very front compartment and hold the door open to prevent the departure of the train, while we last-second arrivals dashed up the length of the train. Seeing us there one Saturday, holding up the departure, and with regard to our high spirits, the driver of the V2 dropped the water-scoop as we sped over Ruislip troughs, in such a way that the water came over the top of the tender – and deluged us completely, as well as a good many more in compartments down the length of the first carriage, I imagine. When we arrived at High Wycombe with spirits now dampened, the driver and fireman were still finding the episode a huge joke.

Another Saturday afternoon, I 'went Southern' to Salisbury, with a 'Merchant Navy' going down and one of the new 'West Country' class engines (now appearing like green mushrooms all over the Southern) coming back, and both of them giving a good account of themselves. I had a love-hate relationship with the design of these particular types of engines, and even today I'm not quite sure if I like them or not, with their air-smoothed casings and Boxpok-type wheels. I remember how we used to refer to the 'West Countries' as 'shufflebottoms' because, instead of having a crisp bark from their chimneys like Great Western engines, they produced a subdued shuffling exhaust note at six 'shuffs' per wheel revolution – 'six beats to the bar', as we called it. We train spotters had a language which was all our own.

Our own new 'Counties' were now being scheduled on the Wolverhampton trains, so I decided to see for myself just how well the newcomers could run. Catching the Saturday afternoon 2.10 to Birkenhead (headed, as it nearly always was, by 6008 *King James II*) I got off at Banbury where I crossed over to the up platform and waited for the 4.06 back to Paddington. It was ten minutes late arriving but sure enough it had No 1012 (later to be named *County of Denbigh*) at its head. We got away in fine style and were soon doing 72 mph at Bicester and seldom did the speed drop below 60 mph, so that by the time we had eased our way through the curves at High Wycombe we had made up for lost time and were running a minute early. Gerrards Cross (according to the log I made of the journey) we took in our stride at 75 mph, then reeled off the rest of the run at cracking pace – until we reached Westbourne Park where signals brought us to a dead stop. Even so, we rolled into Paddington two minutes early.

A friendly-looking driver was leaning out of the cab as I walked past No 1012 as it simmered at the buffer stops, so I commented on the fine run and asked him what he thought of the new engines.

'Good runners,' he replied. 'A bit unsteady on their feet though.'

Like all outside-cylindered two-cylinder engines, the 'Counties' were inclined to 'waddle', probably more so because of the high pressure at which they worked. This, together with the heavy hammer blow they dealt on the rails, had much to do with their working pressure being reduced to 250 psi later on. By and large, they were never a particularly popular engine with either enginemen or the civil engineering department, but they could perform well enough as I had witnessed, and I became rather fond of the 'Counties'.

It doesn't do to speak ill of a Great Western engine, but I do recall one morning seeing one of these 'Counties' stuck fast. It was brand new, in pristine condition and still without a name. It had backed on to its train quite normally but when the 'Right away' was given it whistled – then refused to budge. No amount of coaxing by the crew would shift it. The

7000 Viscount Portal, one of the post-war 'Castles', leaving Paddington with the Torbay Express, August 1958. The parcels platform, where I would watch the 'Limited' go by, is on the right, and above the bridge is the office block which housed the dining club on its top floor and which offered a panoramic view all over the station approaches (*Brian Morrison*)

The LNER's contribution to the early post-war locomotive resurgence scene: rebuilt Pacific 4470 *Great Northern*, still in the short-lived deep blue livery lined with red which it wore at its Marylebone debut, starting out from King's Cross on the 10.30 a.m. to Leeds in December 1945. Soon afterwards, it was given a full-depth cab and smoke-deflectors and painted in standard LNER apple-green (*C. C. B. Herbert*)

engine had come to rest 'on centre', with one of its piston rods fully extended so that this and the connecting rod were in line and the crank pin on the wheel at its furthest point away from the cylinder. In theory, being in a 'stuck fast' situation should never arise with any two-cylinder engine as, with one piston 'on centre', the other one is half-way along its stroke and capable of producing maximum power to get the engine moving. But not this engine. It defied all reason, and the frantic efforts of its crew to get itself going. Engulfed in clouds of steam from the safety valve which was now blowing hard, and the cylinder cocks which were open for starting, it sat there, motionless – until a nudge from the rear by the pannier tank which had brought in the 'empties' shifted it 'over centre', and away went the 'County' in fine style as if nothing had happened.

Just what caused that hiccup, I have tried to work out many times – even discussing it with enginemen. The theory is – and dare I say it? – that Swindon hadn't set the valve gear quite right and the valves were a bit out on their spindles. There must be another reason though.

Another new engine to come on the scene was the oil-burner. Standing in line for lunch in the staff dining club (the canteen) on the top floor of the tall goods office building in Bishop's Road, I had a view all over the approach to the station, where I caught sight of another engine with an unusual tender. Forsaking the prospect of lunch, I dashed down to the platform where I found 'Castle' 5091 *Cleeve Abbey*, restored in pre-war finery and equipped with one of the smaller 3500-gallon tenders, on top of which and perched where the coal should have been, was a big square tank for supplying fuel to the oil-burning equipment with which the engine had been fitted. Chatting to the fireman as he waited for the 'Right away', he told me that he much preferred 'twiddling knobs' to heaving coal, and thought it would be a good thing for firemen if all Great Western engines were converted to oil-burners.

I saw quite a few oil-burners after that, including another 'Castle', 5039 *Rhuddlan Castle*, as well as a number of 'Twenty-eights' and 'Halls', all converted because of the acute shortage of coal so evident in that period just after the war. Coal-burning engines were having to put up with some very inferior fuel, evidence of which could be seen in the smoke-screens they laid across Paddington as they digested briquettes, some square, some egg-shaped, made from compressed coal dust – while their own favourite fuel, Welsh steam coal, was being exported as fast as it came out of the ground. We agreed at the time that it did seem a little odd that we should be selling coal abroad and, with the dollars earned, buy oil fuel to run the trains which should have burned the exported coal in the first place.

Brand-new 'Castles' started appearing – the first one I remember seeing at Paddington was 7000 *Viscount Portal* – looking exactly like their pre-war counterparts. Nearly everyone in our department took an interest in all things new on other railways as well as the Great Western, particularly locomotives, and the building of new 'Castles' caused one or two raised eyebrows. Whereas the GWR was happy to continue to lean on a design dating back to 1923, other railways were forging ahead with new and advanced designs, particularly the Southern who were making great play with their new lightweight 'West Country' Pacifics. The LNER, too, had brought forth their new Pacific (albeit a rebuild of an older engine, nevertheless a *complete* rebuild), 4470 *Great Northern*, which I had found on display in royal blue splendour one evening at Marylebone station as I headed homeward. So – where was our new Pacific? Here, Huddy and I joined forces, I recall, and held forth that a 'Castle', given the right sort of coal, was an extremely efficient, economical and compact power unit, as good as the rest of them and any changes going on were under the surface in the form of improved superheating, etc.

Far from stagnating, the Great Western was looking further ahead than other railways and was investigating an entirely new form of traction, and had ordered a gas turbine locomotive from Switzerland (shades of the de Glehn Compounds all those years before, when in 1903 the Great Western had 'bought French'). This led to conjecture (as well as doodles) in the office as to how this new engine would actually work as, at the time, anything to do with gas turbines was associated with the recently introduced jet engines on aircraft and this conjured up all sorts of amusing visions (and doodles) of jet-propelled trains. Two gas turbine locomotives were eventually purchased; one, Swiss, No 18000, came from the Brown-Boveri Company, and the other, British, No 18100, from Metropolitan

'Castles' were still further developed after GWR days. 4074 *Caldicot Castle*, second of the 'Castles' to be built in 1923, seen here in the early 1960s much modified by BR, having higher-superheat boiler, double chimney, mechanical lubrication, easier-curve steam pipes and later-type inside cylinder block, as well as a flush-sided tender (*R. C. Riley*)

The shape of things to come, had the Great Western survived? The first gas turbine locomotive, 18000, built in Switzerland by the Brown-Boveri Company, outside Swindon works, October 1954 (*Brian Morrison*)

Vickers. They were not jet-propelled, of course, but had gas turbines which drove generators to supply electricity to traction motors coupled to the wheels. They were delivered after the Great Western Railway had been nationalised and, although evaluated, the gas turbine was considered by British Railways as not being a suitable (or economical) method of traction.

Looking still further ahead, plans which had been made earlier for the electrification of the main line were brought out, dusted down and discussed at length. The Divisional Engineer's Office became involved in the possibility of future electrification when it was decided to raise one of its bridges some four inches to give clearance for overhead wires, should they ever be installed – not a complete bridge, but a structure over the main line west of Maidenhead which had been started just before the war and left as a skeleton of girders throughout the war years – a favoured train-spotting place for me in my schooldays. My own feelings are, on the strength of what I saw and heard, that the Great Western, had it not been nationalised, would most certainly have 'gone electric' in time.

And I bought a brand-new copy of *GWR Engines: Numbers, Names, Types & Classes* on the station bookstall one morning, only now it was no longer a 'shilling book for boys of all ages' as it cost 2s 6d – and had just as much appeal. Once a train spotter, always a train spotter.

King Steam was to stay in power for another twenty years, yet even at that time I detected a shift of heart in the Great Western's publicity material. Until then, it had always championed steam as its main source of traction but now, in my new GWR engine book, there were seeds of doubt, with more than just a hint that familiar things such as crossheads and connecting rods would disappear, along with the plume of smoke and the familiar shape of the locomotive. Oil, and the gas turbine locomotive, was to change everything, it was said; all of which I found more than a little disturbing, especially as it came from a source I had trusted through the years. Yet, on reflection, I accepted it reluctantly. Technical advances in other fields had gone ahead by leaps and bounds during the war. Now, I realised, it was the turn of the railways.

During my time on the Great Western, my interest in locomotives had gone way beyond that of being just a train spotter. My doodling of locomotives had become accepted but never again allowed to get out of hand – Huddy saw to that – and my enthusiasm must have been obvious as, at one time, feelers were put out on my behalf to see if I could join the Locomotive Department. Wartime restrictions on labour still applied, however (even after the war), and as it was considered that I was of more immediate use to the Divisional Engineer, I had to stay put. On reflection, if I had gone over to mechanical engineering, I would not have had that wonderful opportunity of roaming round the System in the way that I had been allowed to do. Old Oak Common, Southall, Slough, Reading, Didcot, Oxford – I had my fill of all the locomotive depots, as well as observing engines at rest at Paddington, engines running free on the main line and engines pottering along branch lines.

I did not stay to see the Great Western Railway become nationalised. When I left, the twin shields of the cities of London and Bristol still blazed on tender sides (along with my initials) and on the waists of coaches. Already, though, with nationalisation in the air, there was a feeling abroad of uncertainty. Jack Rees, the trade unionist, on the other hand, was all for it – and told me so – many a time. I hoped, as others did, that they would keep it as the Great Western Railway after nationalisation and we saw no reason why the old institution should not survive, if only in name, as it had done for over a century, weathering the storm of grouping to form the four big companies in 1923 on the way. The Great Western couldn't disappear – it belonged – it was a survivor. But, of course, it did – and the powers-that-be spent thousands of pounds on thousands of gallons of paint to try and cover up the traces of the Great Western with new and outlandish colour schemes, but as long as the old loyalties and its steam locomotives remained, the Great Western shone through the layers of paint.

The last job I did for the Great Western was a lock-up store for valuable parcels in transit, and many were to be the times when I was heading out of Paddington for the West Country in later years that I would glance across the lines towards the goods platform at Westbourne Park at this not-very-imposing structure and think to myself, 'I did that.' Reggie handed me this job which was to be my swansong and left me to get on with it. I saw it through from beginning to end, from discussion on the site, measuring up,

In the late 1950s, credit must be given to BR for restoring many of its trains to their former glory. Green and chocolate-and-cream combine once more as 6000 *King George V*, looking much as it does in preservation today, heads the Cambrian Coast Express out of Paddington in August 1960 (*R. C. Riley*)

The setting for my swansong: 2-6-2 tank engine No 6148, having just left Westbourne Park station en route for Paddington on 30 November 1957, passes the goods platform where I built my lock-up store. Flat-bottom track is evident on the near line next to the train, the 'down main' (R. C. Riley)

design on the drawing board, ordering the material (angle-iron, wire mesh – even a padlock and keys) and keeping an eye on its construction. Ironically, my final contribution to the Great Western Railway, as it was about to release me, was – a lock-up.

Along with nearly all the young men of my generation, I joined the Forces – the Royal Navy. With so many men having drifted back from war service, the Great Western was well up to strength, and it was decided that individuals such as myself could now be released to serve our country. All of us – we all had to go at some time or another – so I went, before nationalisation set in.

Ask me what I would enjoy doing most if I could have some of my Great Western time again and the answer would be easy – and certain. It wouldn't necessarily be a trip on the 'Limited', and heaven knows, I'd love to watch the 'Limited' go by again. No, my choice wouldn't take up much more than half an hour of the past, and would be something I never ceased to enjoy, time and time again. A trip on a train to Paddington from Reading, that would be my idea of heaven – steam-hauled, of course – a 'fast', with preferably a 'Castle' or an old 'Star' up front (though a 'Hall' or one of the old 'Saints', which could really lope along, would do just as well), with me in a corner seat facing the engine in one of those low-waisted-stock compartments with the big windows and near enough to the engine so that I could hear the ATC bell telling me we had a clear road ahead. Wintertime – a late afternoon in December with the dusk outside frosty and the air crisply clear. No matter if the compartment fills up, as I would be lost – completely lost – in the pure magic of that journey.

The sharp-blast getaway, the gathering speed into the dusk, familiar stations flying past, the lights of London coming up, and a clear run all the way, not even a hold-up at Westbourne Park – to ease gently round into the station and come to rest at the 'stops' within the wintry atmosphere of Paddington station – bang on time.

Mine, though, is not such a remote dream after all, as I have spent many a happy hour in recent years, travelling behind steam trains. Thankfully, so many steam engines managed to escape the cutters' torches and survive today – from little tank engines to giant main-line locomotives, and surely the interest in King Steam has never been greater. I, for one, can admire the band of stalwarts who undertake such marathon tasks by putting them in such first-class condition to allow such as me to wallow in our corner seats of nostalgia. A pity that one of my 'Counties' didn't survive, or perhaps even a 'Bulldog', but with so many examples of all the railway companies in preservation there are plenty more favourites to choose from, and each steam-hauled train becomes a thread of glorious history running back into the past from this microchip age.

No amount of nostalgia, though, could ever recapture for me that atmosphere I would ask to meet at Paddington – the lamps shining through the hazy gloom, people hurrying homewards and stepping aside to dodge the seemingly endless trains of tractor-hauled luggage trolleys (at least one of which seemed always to limp noisily with a faulty tyre) – and that aroma – that indefinable combination of steam locomotives, restaurant cars, coffee, warm air wafting up from the Underground, carthorses – and even fish, which always seemed to travel by passenger train. And the whole scene punctuated by the shrill blasts of whistles, and the friendly voice of the station announcer as, with confident dulcet tones, she gave the destinations of a whole plethora of early evening trains. She always knew where they were going, those trains – back along the way I had just come, westward and to the Midlands, and each one headed by that most noble of beasts devised by man – a steam locomotive.

Come to think of it, I never did get around to meeting the station announcer and seeing where she worked. I wanted to, and Reggie often said he would take me there but somehow he never got round to it. He knew her, of course.

It was Fred Cockhill, though, who had the last word. Apparently I wasn't the first offender at drawing-board doodling. Coming up to my board just before I left, Fred cast his coroner's-inquest eye over my backing sheet and said, 'You know, we used to have another young lad here before you – now he really *could* draw railway engines.'

I like to think that, after I left, a youngster sat at my drawing board and doodled, and that Fred said the same thing to him – only this time he was referring to me.

Still going strong: 3440 *City of Truro* pulls away from Northwood towards Bewdley after its full restoration by the Severn Valley Railway for the 150th anniversary of the Great Western Railway in 1985. One of many engines which have provided me with 'a corner seat of nostalgia' (*Tony Dyer*)

LOCOMOTION

GREATEST among the first railway engineers in the early years of the nineteenth century was George Stephenson, a self-educated man of humble origins who had set himself up both as a builder of railways as well as the locomotives to run on them. Gaining experience from his first locomotive in 1814 whilst employed at Killingworth colliery, George Stephenson went on to produce other engines, each progressively better, before setting up a locomotive manufacturing business in 1823 with his son Robert and two other partners, thus forming the famous firm of Robert Stephenson & Co in Newcastle-upon-Tyne – soon to build *Locomotion* and eventually to blossom forth to supply the world with its locomotives.

Built at a cost of £500, *Locomotion* was delivered to the Stockton & Darlington Railway (claimed as the world's first steam-operated public railway) eleven days before it opened on 27 September 1825, to be the only steam engine employed at first on an otherwise horse-drawn railway. Weighing a mere $8\frac{1}{2}$ tons, *Locomotion* was regarded as being the best-built and most advanced locomotive of that time, though in truth it was little more than a boiler mounted on four wheels, on top of which were carried the cylinders and motion, all of which bore a strong resemblance to stationary steam engine practice of the day, even to the parallel motion, arranged to support the piston rods throughout the length of their travel.

Balanced on the ends of the piston rods were two long cross beams, each carrying two slender connecting rods reaching down to couple the motion to the crankpins on the wheels. With cranks on the leading pair of wheels set at 90 degrees to the trailing pair, thrust was evenly established on the wheels, as well as providing ease of starting as one pair of cranks was always in a position to receive piston thrust. Whilst in motion, the whole configuration with its exaggerated up-and-down movement was said to be likened to two grasshoppers chasing each other.

The two cylinders of 8-in. bore and 24-in. stroke were half submerged into the boiler, which in turn was of the 'Cornish' pattern (again, a type in current use with stationary engines) having a single large tubular flue in which the fire was laid and which reached throughout the whole length of the lower half of the boiler to the base of the chimney. Good steaming capabilities still eluded those early locomotive pioneers though George Stephenson had gone a long way to improve matters on *Locomotion* by introducing exhaust steam from the cylinders into the chimney where it created a draught through the flue to liven up the fire and increase steam production.

Firing the boiler was achieved by a man who stood at the front end of the platform of the oak-built tender, while the driver took up a precarious position on the footboard alongside the motion (on the far side in the illustration) where he operated the regulator-valve handle and effected the reversing of the engine – a deft operation which required the disengaging of two levers (forming part of the valve gear linkage) and using them to alter the valve positions in the cylinders to cause enough movement of the axle for stops to come around and engage the loosely fitting eccentrics and engage them in their new position to give valve movement for the altered direction of travel.

Locomotion remained active on the Stockton & Darlington Railway for twenty-one years, after which it took up residence – and still does – as a permanent showpiece at Darlington Top Bank station, coming down from its pedestal on a number of occasions to be exhibited as far afield as the USA (twice) and Paris, as well as visiting exhibitions at various places in the United Kingdom, including the British Empire Exhibition at Wembley in 1924. At one time, *Locomotion* suffered the indignity of being petrol-driven when an internal combustion engine was concealed in the tender to allow the engine to take an active part in the Railway Centenary Celebrations of 1925.

The 150th anniversary of the Stockton & Darlington Railway (for which this portrait of *Locomotion* was commissioned) was celebrated in style in August 1975 when a grand cavalcade of locomotives of all shapes, sizes and ages paid homage to George Stephenson, his railway and his famous engine. Leading the parade was a full-size working replica of *Locomotion*, built under the direction of engineer Mike Satow, using both modern techniques (the boiler had to be welded) and traditional skills – so much a part of the industrial heritage of the north-east corner of England – in its making. This replica *Locomotion* now has its home at the North of England Open Air Museum at Beamish, not far from Newcastle-upon-Tyne, and during summer months may be seen in action, displaying its own peculiar brand of 'poetry in motion'.

Stockton/Darlington Rly *Locomotion*

ROCKET

THE story of the celebrated little engine *Rocket* and how it achieved fame in competition with three other locomotives at the Rainhill Trials is a legend unto itself in the annals of English history. Undecided whether to use locomotives or stationary engines using ropes to haul its wagons, the directors of the Liverpool & Manchester Railway had offered a prize of £500 for the 'Most Improved Locomotive Engine', laying down certain specifications which, if met, would convince them that locomotives should provide the motive power. These included a ceiling cost of £550 to build, a maximum weight of 6 tons, and a stipulation that the engine must consume its own smoke. This smoke consumption was a law of the day, but exactly how it was to be achieved seems never to have been explained, though the use of coke as fuel in early locomotives went a long way to satisfy the law as it burned with little or no smoke.

The trials, lasting eight days and watched by some 15,000 spectators, took place at Rainhill near Liverpool in October 1829. The outright winner was George Stephenson's *Rocket*, designed by him specifically for the contest and built at the works of Robert Stephenson & Co at Newcastle-upon-Tyne in the four months before the event took place.

The success of *Rocket* was due in no small measure to the design of its boiler in which George Stephenson now used a multiplicity of tubes instead of the more usual large single-flue arrangement in common use at the time. Not an entirely new idea, the suggestion to do this had come, not from another engineer, but from the Secretary and Treasurer of the Liverpool & Manchester Railway, a certain Mr Henry Booth. Using 25 copper tubes of 3-in. diameter between firebox and chimney, and combining the arrangement with a forced draught brought about by exhausting steam from the cylinders into the chimney, Stephenson had devised a system which not only provided *Rocket* with all the steam it needed, but one which was to become standard practice in boilers throughout the whole of the steam era.

Unlike later practice, however, the firebox of *Rocket* was not an integral part of its boiler but a separate structure of sheet copper bolted on to the back, water-jacketed on the top and sides with four copper circulating pipes connecting this water space to the boiler. Steam pressure was 50 psi, this being indicated by a rod protruding from a long tube reaching upwards alongside the chimney, the lower half forming a deep hairpin U which contained mercury, to be displaced by steam pressure and push the rod further out of the top of the tube as pressure built up – a primitive but effective way of indicating pressure in the days before the pointer-and-dial pressure gauge had been invented. No glass water gauges in those days either, but two brass try-cocks (forward of the nameplate) to test the level of water in the boiler.

Two cylinders of 8-in. bore and 17-in. stroke sloped downwards at an angle of 37 degrees and were connected to the iron crank bosses of the wooden driving wheels, set at 90 degrees to each other, with a crosshead-and-slidebars arrangement now replacing the parallel-motion system used on *Locomotion*. Reversing was achieved in much the same way as on *Locomotion* though, but now a foot pedal on the footplate was depressed to release a dog clutch on the driving axle and allow the eccentrics to turn freely.

Having won the day, *Rocket* continued to work on the Liverpool & Manchester Railway until 1836, during which time the cylinders were moved downwards to a new position a few degrees above horizontal and a more conventional smokebox fitted, supporting the chimney on its top. In this condition – though somewhat dilapidated – it eventually reached the Science Museum in London – though not before *Rocket* had worked for a few years on the private railway serving Brampton colliery in Cumberland, to be laid aside derelict at some time around 1840.

The remains of the original *Rocket* reside in the Science Museum today, along with the full-size replica of the engine (sliced away on one side to show its workings) on which this illustration of *Rocket* is based. This replica owes its existence to Henry Ford, the American motor car magnate who commissioned Robert Stephenson & Co to build for him a full-size exact working replica of *Rocket*, a century after the original was born, in 1929. This was no mean task as no proper drawings had ever been made and a good deal of alteration had taken place over the years. By a tremendous amount of painstaking research and detective work the details were pieced together and the replica *Rocket* built, along with two more for American museums during 1931, and the fourth for the Science Museum in 1935. A full-size working edition of *Rocket* was built to celebrate the Rainhill Trials in May 1980, again by that master-builder of replicas, Mike Satow.

Stephenson's *Rocket*

GNR 'STIRLING SINGLE' No 1

and the '8-ft Singles'

THE name of Patrick Stirling will be forever linked with the distinctive locomotives he designed and which once ranged over the Great Northern Railway between London and York. Stirling, a Scotsman, had joined the Great Northern at Doncaster in 1866 and his brief had been that of economy, both in the cost of building engines and their operation in traffic – and in their design, too, for the sheer economy of line he implemented led to their elegant, and distinctive, shapes. With their 'straight-back' boilers (having no dome), single large driving wheels and somewhat plain cabs backed up by large tenders (which did much to add to their air of purposefulness) the design of the 'singles' came together to produce Stirling's own unique brand of locomotive, each to be identified by a number as none carried a name.

No 1 was built in 1870, and whereas other smaller-wheeled 'singles' designed by Stirling carried their front ends on one axle, with No 1 he employed a bogie. Not in the least sophisticated, it had none of the side-to-side control utilising suspension links or the crosswise spring arrangements of later years (by which the bogie eased the weight of its locomotive into a curve), but was simply pivoted by a single pin some 3 in. back from its centre point, a system which appears to have been perfectly adequate.

Outside cylinders were employed to drive the huge 8 ft 1 in. diameter driving wheels (lending the name of '8-ft Singles' to the class), the decision to place them there being influenced by the need to keep the boiler as low as possible, as cylinders fitted within the frames would have meant raising the height of the boiler to give clearance for a cranked axle. Outside cylinders were not a new feature – after all, *Rocket* had used them.

Not entirely successful at first, No 1 had its frames lengthened in 1877 to accommodate a new and longer firebox within its boiler, and in 1880 was completely rebuilt, its frames being strengthened and a new boiler fitted – which is the condition in which No 1 is shown here. In the meantime, new 'Singles' were being built, in pairs, to the similar pattern of No 1 with each successive pair having minor differences and receiving random numbering as it was Great Northern policy to perpetuate numbers on new locomotives as older ones were withdrawn.

Altogether, fifty-three '8-ft Singles' were built over a period of twenty-five years, each a worthy performer despite the ever-increasing weight of express passenger trains. For a long time, no vacuum brake pipe appeared at the front end of a 'Single' as Stirling was adamant that his locomotives were adequately powered for any Great Northern train, and it was only toward the end of his regime that in 1894 he reluctantly agreed to double-heading with his 'Singles' and they sprouted vacuum pipes above their front buffer beams for that purpose.

Possibly as much has been recorded about No 1 in retirement as during its active life. Taken out of service in 1907, No 1 was brought out two years later to be displayed at the 1909 Imperial International Exhibition at White City in London, to be seen as a comparison with one of the Great Northern's latest engines, 'Atlantic' No 1442, after which it was lain aside in King's Cross sheds. No proper tender was available for this appearance so a less heavy-looking one was found (actually, pre-dating No 1) and is the one fitted to the engine today. In 1921, No 1 was brought out and dusted down to pose in publicity photographs alongside Gresley's new three-cylinder 2-6-0 No 1000, and the following year it posed again, this time with another new Gresley engine, the second of his 'Pacifics', No 1471. In store once more, the next emergence came in 1925, on this occasion in steam to take part in the Railway Centenary Celebration, going afterwards into the York Railway Museum.

An excuse to bring out No 1 again was looked for – and found eventually – on the 50th anniversary of the 1888 Race to the North (see Caledonian Railway No 123) in 1938, when Sir Nigel Gresley himself was approached to see if he would allow the old Stirling 'Single' to run at the head of a train made up of typical carriages of the 1880s to mark the event. Gresley gave the enterprise his blessing and once more, following a quick overhaul at Doncaster works, No 1 took to the rails to run excursions from King's Cross and various towns and cities on the LNER system.

Two noteworthy occasions since then have been honoured with the appearance of the old Stirling eight-footer, namely the Stockton & Darlington 150th anniversary at Shildon in 1975 and the *Rocket* celebrations of 1980. Back now in the National Railway Museum at York, it is to be hoped that there will be further opportunities for the venerable Stirling 8-ft 'Single' No 1 to display its graceful lines and easy big-wheel action on the main line again.

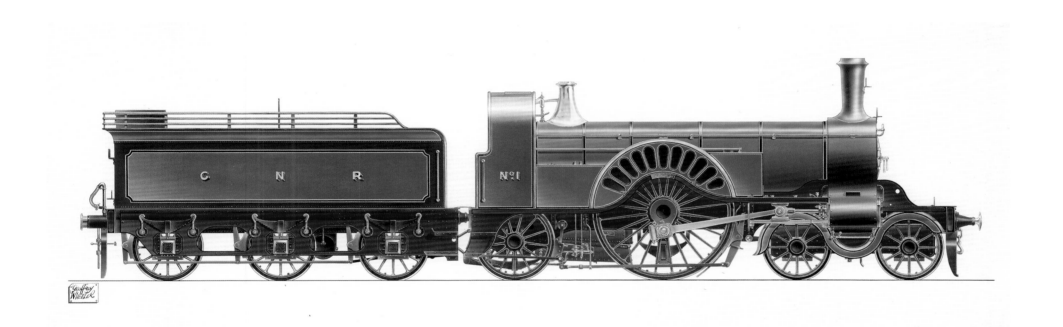

GNR 'Stirling Single' No 1

WHAT must have been the most famous railway race in history took place in the summer of 1888, when the two main lines northward out of London tussled with each other for the fastest trains to Edinburgh. Almost daily throughout those summer months the pace of the northbound trains quickened and timings were lopped – from 10 hours to $8\frac{1}{2}$, then down to just over $7\frac{1}{2}$ hours until, in August, a truce was called as things were getting out of hand. Although the East Coast Route out of London (King's Cross) had gained honours for the fastest timing (snatched on the very day the truce was being discussed), it was an engine on the West Coast Route to whom laurels were really due, as it had the distinction of being the only one used without a break throughout the whole escapade. This was the Caledonian Railway's No 123, the engine which took over the train from London (Euston) when it reached Carlisle then, every single day of the races, hauling it at a truly cracking pace, onward to Edinburgh (Waverley) – a 101-mile journey which it covered in just over 100 minutes, taking in its stride the 10-mile laborious climb up and over Beattock summit on the way.

Whether or not the Caledonian Railway had foreseen the forthcoming contest (which was very unlikely) when two years previously they had ordered the engine – or indeed, whether they had any call at all for a bogie single at the time when engines such as their Drummond-designed 4-4-os were filling the bill by doing everything expected of them on express work – may never be known, but with No 123 the Caledonian certainly had the right machine waiting in the slips to play its part to the full in the races. The engine itself was a 'one-off' job, built by Neilson & Co of Glasgow who had wanted a showpiece to represent them at the Edinburgh International Exhibition of 1886, at rather short notice. This resulted in the building of No 123 in record time – in no less than sixty-six days, from towards the end of January when the Caledonian placed its order, to delivery on 1 April 1886. It was delivered on time, too – a cleanly designed, neatly arranged engine, looking every inch a Caledonian and resplendent in the company's blue-and-red livery displaying the Caledonian Railway crests on its tender but, as yet, lacking the attendant CR initials. These it was to receive after the Exhibition, as well as brackets for its forward-facing lamps, to be fixed, Caledonian-fashion, high up on the cabsides instead of the more usual position above the front buffer beam as on other railways. A unique detail was the air-assisted sanding gear for the large 7-ft diameter driving wheels, compressed air being supplied by the Westinghouse brake pump just forward of the cab.

After the dust of the races had settled, No 123 went back to everyday passenger train duties, working at first from Edinburgh on expresses to Carlisle and back; then, stationed at Carlisle, taking a twice-a-day out-and-back express turn to Glasgow and covering 409 miles daily. It was becoming obvious, though, that with increasing train-loads there was less and less demand for a 'racehorse' whose speciality was dashing along with a lightweight train of just four or five coaches in tow, although its usefulness as a breadwinner was extended for a while by sending it north to Perth to work on the Aberdeen trains. In the course of time, however, No 123 was relegated to even lighter – yet somewhat distinguished – duties, that of taking the Caledonian Railway directors' saloon around the system. During the course of its forty-nine years' active service, though, an even more distinguished duty came the way of No 123 – that of Royal Train Pilot, to travel on ahead of the royal train whenever it covered Caledonian metals, and this it did during the reigns of three monarchs: Queen Victoria, King Edward VII and King George V.

Caledonian No 123 has pride of place in railway history, too, as it was the last main-line engine with single driving wheels to remain in revenue service in Great Britain. When the Caledonian was absorbed into the London Midland & Scottish Railway in 1923, No 123 continued to work, only now it received the number 14010 and lost its Caledonian livery for one of Midland red. In 1927, a 'new' boiler became due (the one used, in fact, was secondhand), which is the one it carries today, with safety valves moved from the top of its dome to a position ahead of the cab.

The decision to preserve this lovely old engine was made on its retirement in 1935 with over 750,000 miles to its credit, when once again it was restored to its original Caledonian blue. Thus adorned, it came out of retirement for a while in 1958 to take to the rails again under its own steam to work 'specials'. Less active nowadays, the old 'Caley 123' resides in the Museum of Transport in Glasgow, a reminder of the days of the graceful 'singles' and a tribute to good, sound Scottish engineering.

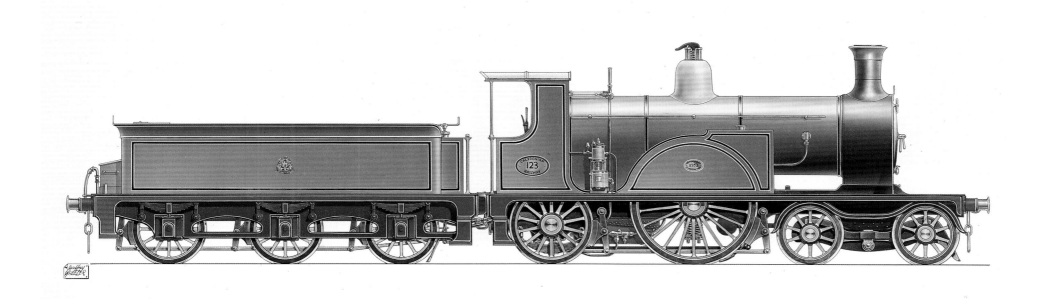

Caledonian Railway No 123

GWR No 3009 FLYING DUTCHMAN

and the 'Dean Singles'

ACCLAIMED by many as being 'the most handsome engines ever built', the class of eighty locomotives to which *Flying Dutchman* belonged were the mainstay of the Great Western Railway's express trains to the West of England in the years immediately before the turn of the century.

Designed by William Dean, they became known fondly as the 'Dean Singles', a reference to their large 7 ft 8½ in. diameter single (as opposed to coupled) driving wheels, the first batch of thirty engines, numbered 3001–3030, being built during 1891–2 as rigid-framed 2-2-2 types with their front ends carried on a single pair of wheels. These included 3009 *Flying Dutchman*, which emerged from Swindon Works in 1892, as yet un-named.

The first eight of these engines to be built, numbered 3021–3028, were constructed as 'convertibles' with wheels outside their frames, set to run on the broad gauge (7 ft 0¼ in.) rails until the final changeover to the standard gauge of 4 ft 8½ in. took place in the spring of 1892, after which they were brought in line with the rest of the batch, with wheels within the frames to suit the new gauge – still as 2-2-2 types.

The weight of these engines (almost 44 tons fully laden) appears to have been too much for a single pair of leading wheels, and it was decided to replace them with a bogie, the final decision to do so probably being made as a result of one of the engines, 3021 *Wigmore Castle*, suffering a broken front axle and becoming derailed in Box Tunnel in September 1893. Frames were extended forward to accommodate the new bogie, and with the front weight now carried on four wheels instead of two, their appearance was completely transformed. Instead of being rather plain, somewhat stubby engines, they took on a new air of graceful dignity balanced with a degree of thrusting elegance, which had been bestowed upon them almost by accident.

Flying Dutchman received this treatment in 1894 and in November of that year emerged once again from Swindon as she is shown here.

In the meantime, production had begun on a further batch of these 4-2-2 engines, the first being 3031 *Achilles*, and between 1894 and 1899 fifty were produced to bring the total number of 'Singles' up to eighty.

Outwardly, the *Achilles* class were almost identical to the modified original engines, but there were minor differences. Trailing-wheel springs, formerly hung below the frames (as on *Flying Dutchman*), were re-positioned above the running plate, and the sloping water-feed pipes mounted alongside the boilers were replaced with vertical pipes when injectors replaced crosshead-driven water pumps on the last engines to be built. As overhauls took place, all the engines were more or less brought into line as pumps were discarded in favour of injectors, and springs became mounted in the overhead position.

In this condition, they ran for a number of years, until many of the engines were 'modernised' and given new 180-lb boilers having Belpaire fireboxes, top-feed and cylindrical 'drumhead' smokeboxes, as well as wider cabs – a move which added to their performance but destroyed completely the elegant proportions that the 'Singles' had previously enjoyed. *Flying Dutchman* was thus modified in 1911.

Renowned as a class as being spirited performers, little is recorded of the exploits of *Flying Dutchman* herself. However, one member of the class, 3065 *Duke of Connaught*, made a niche for itself in railway history on the occasion when it took over the mail train from *City of Truro* at Bristol after its record-breaking run. Reeling off the rest of the journey to Paddington at an average speed of over 70 mph, 3065 *Duke of Connaught* maintained an average of 80 mph for the last seventy miles and claimed a maximum of almost 92 mph on the way.

During Queen Victoria's Jubilee celebration in 1897, two engines of the class were selected for royal train duties – 3041 *The Queen* (specially renamed from *Emlyn* for the occasion) and 3044 *Empress of India*. Both engines were given individual treatment, being 'double-lined' by having twin orange-black-orange lining-out on their green paintwork. A full-size replica of *The Queen* now resides at the Royalty and Empire Exhibition at Windsor.

Despite re-boilering and its resulting increase in performance, the usefulness of the 'Singles' was short-lived as very soon more powerful four-coupled and six-coupled engines were needed to handle the increasing loads to and from the West Country. Withdrawal of the 'Singles' began in 1908 and by 1915 all eighty engines had gone, and the reign of the single-wheeler that had served the Great Western from its very beginning was over, *Flying Dutchman* itself being withdrawn in February 1914.

GWR No 3009 *Flying Dutchman*

GWR No 8 GOOCH

and the 'Armstrongs'

THE mid-1890s certainly marked a zenith of elegance in locomotive design on the Great Western Railway. It was as if William Dean had 'got his eye in' and was now turning out some beautifully proportioned engines as a result — engines resplendent in paintwork which seemed completely to complement the needs of the pleasing outlines to which it was applied. Following closely on the heels of the handsome 'Dean Singles' in 1894, the four engines making up the 'Armstrong' class achieved a degree of aesthetic balance seldom equalled in a locomotive, to be blessed, appropriately, with the names of eminent personages who had given yeoman service to the GWR: No 7 *Armstrong*, No 8 *Gooch* (shown opposite), No 14 *Charles Saunders* and No 16 *Brunel*.

These engines could well be considered as being four-coupled versions of the 'Singles', which mainly they were, making use of the same 160-lb boiler, bogie and inside motion (having, though, cylinder bores opened up from the 19-in. diameter of the 'Singles' to 20-in.), but having driving wheels reduced to 7-ft diameter, this being the last instance of driving wheels of this size being used by the Great Western. Officially, though, these were not new engines but rebuilds of earlier 2-4-0s, to which the 'Armstrongs' bore only a fleeting resemblance. Both No 7 and No 8 had been experimental compound engines, neither being deemed entirely successful (No 8, a broad-gauge engine, is recorded as having blown three of its four pistons and cylinder covers to smithereens on its maiden voyage), whilst No 14 and No 16 had been broad-gauge 'convertibles', designed to work between Bristol and Swindon and, in their time, the most powerful pair of passenger engines on the GWR. Eventually, the duties allotted to the resulting 'Armstrongs' took them over the same ground as all four locomotives were scheduled on the London (Paddington) to Bristol run on express passenger trains.

The tenders attached to the 'Armstrongs' were the 3000-gallon pattern, the same as those used on the 'Singles', and which employed a simple form of water-condensing apparatus. Exhaust steam could be tapped via a valve and fed back to the tender, where it was led up inside the vertical closed-ended column seen standing proud towards the rear of the tender, to condense as water and run back down into the tender's tank. Not only acting as a condenser, this system offered a degree of feedwater heating — no doubt preventing freezing during cold weather as well. A pair of pumps, attached to the engine's crossheads, driven directly by the pistons, fed water into the boiler via the sloping pipes and large clack-valves seen at the front end of the boiler, backed up by a single live-steam injector beneath the footplate which fed a clack-valve on the boiler backhead inside the cab — handy for topping up the boiler when the engine was stationary.

The two slender columns also seen reaching above the tender each carried an eye at the top through which a cord was threaded, connected to the engine's whistle and passing back along the carriages — the Great Western's form of communication cord in those days before the advent of the automatic vacuum brake.

With their sturdy double-framed layout and coupled driving wheels sandwiched between the inner and outer pairs of frames (the leading pair of driving wheels had four bearings in axle-boxes borne by four sets of springs supporting its cranked axle; the trailing pair were supported by only two axle-boxes and springs on the outer frames), *Gooch* and her sister 'Armstrongs' represented an important advance in Great Western locomotive design as they were to set the pattern for a range of inside-cylinder 4-4-0 tender engines for quite a number of years to come, to include such classes as the 'Dukes', 'Bulldogs' and 'Cities'. The 'Armstrongs' themselves eventually became attached to the 'Flower' class (*Gooch* became No 4172), carrying tapered domeless boilers and fitted with 6 ft $8\frac{1}{2}$ in. driving wheels and, unlike the 'Singles', found themselves a niche of usefulness well into the 'big engine era', working in less-demanding passenger train roles.

With their demise (the last to go was *Charles Saunders* in 1930), Joseph Armstrong, who had been William Dean's predecessor as Locomotive Superintendent, and Charles Saunders, the Company's first Secretary and Superintendent of the Line, were no longer to be commemorated on locomotive nameplates, but the names of Brunel and Gooch, those two stalwarts who had, between them, forged the Great Western Railway and got it moving, were restored to engines a few years later, on 'Castles' 5069 *Isambard Kingdom Brunel* and 5070 *Sir Daniel Gooch*.

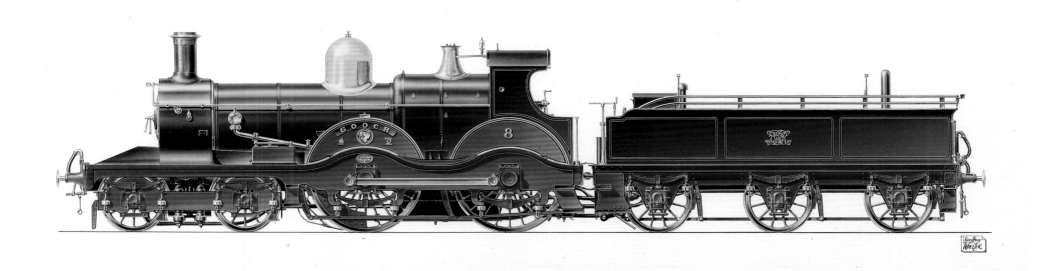

GWR No 8 *Gooch*

GWR No 3252 DUKE of CORNWALL

and the 'Dukes'

MORE commonly known in their day as the 'Devonshires' (or simply the 'Devons'), the 'Dukes' were originally designed to work to and fro over the heavily graded South Devon main line. Newton Abbot was just about the limit of usefulness for a 'Dean Single' on a West of England express train – they were racers rather than hill-climbers – and tackling the hilly terrain beyond was left to the 'Dukes', a task to which they were eminently suited with their smaller 5 ft $7\frac{1}{2}$ in. diameter driving wheels, coupled together for extra adhesion.

Built in 1895, 3252 *Duke of Cornwall* was the first of forty engines, numbered 3252–3291, to be turned out from Swindon over the next two years, and whose usefulness was to extend eventually far beyond the West Country inclines they had been designed to tackle to the whole of the Great Western system, in both passenger and freight train roles. Following the pattern established by the 'Dean Singles' by having double frames with driving wheels sandwiched between the inner and outer frames, inside cylinders and 160-lb boilers with round-topped fireboxes (now flush) and huge brass domes, the 'Dukes' were the first of the new classes to be developed from the 'Armstrongs'.

The original 'Dukes' (certainly the first twenty, including *Duke of Cornwall*) were fitted with Mansell-type wooden-centred bogie and tender wheels, but after a while these were replaced with spoked wheels of a more conventional pattern, suggesting that those with wooden centres had only a limited amount of success. In order to fit on the somewhat smaller-diameter turntables in the territory on which they worked, short 2000-gallon tenders were attached to the 'Dukes' at first, but as turning facilities improved and their usefulness extended further afield, 3000-gallon tenders similar to those of the 'Dean Singles' and the 'Armstrongs' were fitted to many of the class. The elongated smokebox housed a wire-mesh grid spark arrestor, placed crosswise between the blastpipe and chimney orifice, and fitted, no doubt, in an effort to minimise lineside fires caused by hot cinders thrown skywards as the 'Dukes' barked their way up and over those steep Devon banks. Water supply to the boiler was delivered by the same twin-pump and single-injector arrangement used by the 'Dean Singles' and the 'Armstrongs', likewise the same condensing arrangement was used. The re-railing jack, seen between the splashers on *Duke of Cornwall*, was standard equipment on Great Western locomotives in those days.

During 1898/9, the class of 'Dukes' was supplemented by a batch of twenty further engines, these now having 5 ft 8 in. diameter driving wheels as well as improved steam ports within their cylinders, being numbered 3312–3331. As twenty random locomotives from these and the earlier 'Dukes' were selected for conversion into 'Bulldogs' (see *Trinidad* and the 'Bulldogs') so the class was brought back to its original total of forty engines, but now with numbers no longer running in sequence as a result of the gaps. New numbers for the class were eventually allocated, running consecutively from 3252 to 3291, *Duke of Cornwall* being the only 'Duke' not to have its number changed – and this it kept until the end of its days in 1937. Much later, the 'Dukes' remaining in 1946 were renumbered again, this time in the 90xx series.

Originally, all the 'Dukes' bore names – a West Country flavour predominating – but their straight nameplates were replaced in time by curved ones carried over the leading splasher. In time, too, Belpaire boilers, still with large domes, replaced original boilers, and during the course of their careers nearly all the 'Dukes' received full-width cabs which extended outwards to embrace the rear splashers and springs. Two exceptions were 3260 (originally 3261) *Mount Edgcumbe* and 3283 *Comet* (originally 3315, later 9083), both of which retained their narrow cabs throughout their working lives.

Despite all the modifications, the 'Dukes' seemed never to lose their dignity and charm, even when painted plain green and black in later years. For some of them, their usefulness lasted for over half a century – right up until the 1950s. 9084 *Isle of Jersey* (originally 3317, then 3284) and 9089 (originally 3326, becoming 3289 – and losing its name *St Austell* on the way) were the last to survive, both 'Dukes' retiring in 1951.

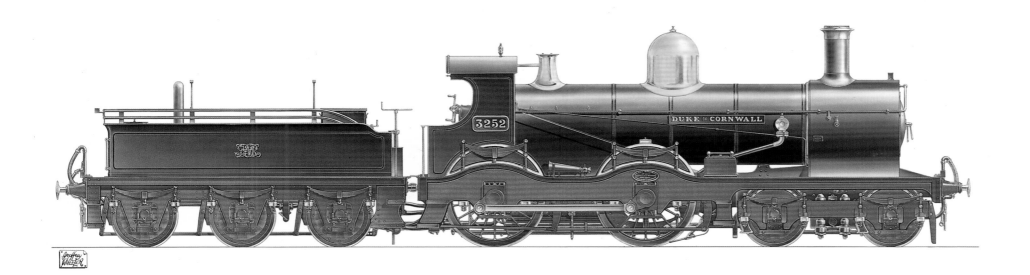

GWR No 3252 *Duke of Cornwall*

NO less than 156 engines similar to *Trinidad* roamed the Great Western system during the early part of the century, and as late as the 1940s there were still more than fifty of them around – enough to make them well worth looking for, although by then they had been relegated to secondary duties. It was an unexpected bonus – serendipity – and a cause for much rejoicing amongst a youthful band of train spotters when one turned up on the main line through Maidenhead, usually on light goods but sometimes on a double-heading turn on an express passenger train.

Engines of the class to which *Trinidad* belonged – the 'Bulldogs' – came in a whole variety of shapes. There were 'Bulldogs' with curved-top frames, others with straight-top frames (some deep, others not so deep), and while many had copper-capped chimneys a lot had those of a plain cast pattern. Some had their sandboxes perched above their frames while others had them hidden out of sight, and whereas the majority of 'Bulldogs' seemed to have coupling rods of the fluted pattern, others carried those of the slab-sided 'fishbelly' type. Throughout the whole class, bogie frames appeared to vary in detail almost from engine to engine, whilst names, too, were even more diverse. *Marco Polo*, *One and All* and *Pershore Plum* were banded together with names taken from the far-flung outposts of the British Empire (*Trinidad* was one of these) and birds, along with the names of towns and dignitaries associated with the Great Western Railway. This assortment of names was carried, in most cases, on conventional nameplates arched over the leading driving wheels, but on many of the early 'Bulldogs' they were displayed on oval combined name-and-number plates mounted on their cabsides. A few had no name at all, these being mostly the ones which originally had borne the names of towns but, it was said, such names confused the travelling public, who assumed that the name on the engine indicated the destination of its train, so they were removed. Tenders varied, too, and could range between the 2500-gallon and 3500-gallon patterns. Shown opposite is one of the 3000-gallon variety.

Originally, the 'Bulldogs', as a class, were never designed as such, but evolved from the 'Dukes', when one of the later batch of 1898/9-built engines was given a much bigger experimental boiler, still with a brass dome (albeit, a smaller one), but now with a square-pattern Belpaire firebox, as well as a much larger cab with a floor-to-ceiling grab-rail across its opening. This was 3312 *Bulldog*, the engine which was to lend its name to the class. The familiar shape of the 'Bulldogs', though, started to emerge with an entirely new 'Duke'-based engine, 3352 *Camel* (named, it may be safely assumed, after the Cornish river, not the beast of burden), which received a boiler – domeless – not unlike that seen on *Trinidad*, having a cylindrical 'drum-type' smokebox mounted on a separate saddle, but with a barrel which was parallel and not yet tapered into the cone-shaped outline which was to become the hallmark of Great Western engines in the years to come. These new boiler shapes were the work of George Jackson Churchward, and although it was to be a couple of years or so before he was to take over officially from William Dean as Locomotive Superintendent (in 1902) he was, with the 'Bulldogs', already beginning to reveal early signs of the tremendous influence he was to have on locomotive design.

As already mentioned, twenty of the 'Dukes' (*Bulldog* included) were selected to join *Camel* and receive these new boilers, then a further twenty new similar engines were built which, along with *Camel*, made a total of forty-one curved-frame 'Bulldogs', to be numbered 3312–3352 and absorbing those numbers vacated by the 'Dukes' in their renumbering. Top of the list was 3312 *Bulldog*, hence the name of the class. New engines were then added to the class, starting with 3353 *Blasius* (later becoming 3341) on which the familiar double-humped curved-top frames inherited from the 'Dukes' now gave way to plain straight-top frames, a feature which was to continue for the rest of the class. No less than a hundred engines similar to *Blasius* were built (these included *Trinidad*), after which came a final batch of fifteen in 1909/10, the 'Birds', these differing again by having more deeply skirted frames, by this time, the taper boiler had become standard.

In time, all the 'Bulldogs' received taper boilers, and as a class renumbered 3300–3455, *Bulldog* itself becoming 3311 and *Trinidad*, originally 3465, taking on the number 3403 as seen here. Following the First World War, all traces of lining out would have gone, and *Trinidad* and her sister 'Bulldogs' would have looked much as she is shown here, in plain all-over green and black, but even such austere treatment could not disguise the fine lines of these rugged little engines – which is how I remember 'Bulldogs'.

GWR No 3403 *Trinidad*

GWR No 3440 CITY of TRURO

THE legend of *City of Truro* and its exploit to become the first British locomotive to be recorded (albeit, unofficially) at a speed in excess of 100 mph in May 1904 has been the subject of more debate than any other achievement in the history of the steam locomotive. Ever since the facts were made known in railway circles through journals such as the *Railway Magazine* and the *Great Western Railway Magazine* by their recorder, Charles Rous-Marten, shortly after the run by *City of Truro* took place, and mention made of a speed of 100 mph being achieved, so the topic has been discussed – often dismissed and even ridiculed. Yet it was not until eighteen years later in 1922 that the Great Western Railway chose to recognise officially the performance of their *City of Truro* (for fear of displaying recklessness and alarming their travelling public, perhaps?) and claim the exactly calculated speed record figure of 102.3 mph as being theirs.

Whatever mystery surrounds the reluctance by the GWR to acknowledge the record and thus denying itself the prestige such publicity may have afforded, there can be little doubt that *City of Truro* did reach – if not exceed – the 100 mph figure. Proof of the excellence of the 'City' class engines had already been made during the previous summer when 3433 *City of Bath* had taken the Prince and Princess of Wales on a truly royal journey from Paddington to the Duchy of Cornwall, to reach Plymouth at an unprecedented average speed of 63.4 mph, and which had included the climb over the South Devon banks on the way, as well as an easing of speed so that the passengers might partake of the royal luncheon in comfort.

The opportunity for *City of Truro* to show its paces had come with the docking of the North German Lloyd liner *Kronprinz Wilhelm* at Plymouth and the need to get its transatlantic mail – including gold bullion – to Bristol and London with all haste. Fierce competition for the rail traffic associated with ocean liners putting into Plymouth was being encountered with the London & South Western Railway, so speed was of the essence if the GWR was to retain its share of the traffic on its longer route to London (via Bristol) as the L&SWR was proving itself no sluggard with its West of England trains.

With Rous-Marten (an eminent and well-respected railway authority) on board the five-van mail train as invited guest and only passenger, to take stopwatch readings between the lineside quarter-mile posts, the record-breaking speed was 'clocked' by him on the descent of Wellington bank from Whiteball tunnel. Here, a timing of 8.8 seconds over a quarter of a mile (producing the 102.3 mph) was recorded, only to be followed by the immediate application of brakes caused, according to Rous-Marten, by platelayers lingering on the line (and here lies the root of any controversy) so that the following quarter-mile's timing could not reflect the speed of the record-breaking quarter and substantiate its claim. Nevertheless, the preceding quarter-mile reading had been 9.2 seconds, and this, added to the record-breaking figure gave exactly 100 mph over the half-mile section – proof enough, the Great Western felt, to validate Rous-Marten's claim.

There can be no doubt that the 'Cities' were magnificent engines, even if they did lack the finesse of appearance which the elegant 'Armstrongs' displayed and from which they sprang. With 6 ft 8½ in. diameter driving wheels and making use of the straight-topped later-type 'Bulldog' frames, the ten 'City' class engines built in 1903 were, in fact, an extension of the 'Atbara' class of 1900, these having parallel boilers and bearing names reflecting the glory bestowed upon places and personalities associated with the Boer War. By equipping the 'Cities' with larger half-cone taper boilers, Churchward had achieved what he had been aiming for – a fast passenger engine capable of sustained steaming. Whereas other railway companies had locomotives capable of bursts of high speeds, GWR engines could now maintain such speeds without running short of steam. Despite this, however, the use of the 'Cities' as top-link performers was comparatively short-lived as the six-coupled express engines were soon to arrive on the scene to take over their roles on the heaviest Great Western expresses.

It may be seen that the locomotive shown here differs greatly from the preserved *City of Truro* so much in evidence today and purists may claim that this preserved, much-modified engine is not the *City of Truro* which broke all records back in 1904 – nor indeed, that in today's condition should it carry the Indian red framing as the rebuilt *City of Truro* wore only lined-out green and black. So be it, but it is the spirit of the old *City of Truro* which lives on, and long may its exploit be analysed and discussed.

GWR No 3440 *City of Truro*

LA belle dame! Whereas, after the turn of the century, British engines (and in particular, Great Western engines) seemed set on drifting away from their Victorian-age elegance, the French – in this instance, anyway – appeared to be intent on retaining it to some degree. Strictly speaking, though, it was an Anglo-French relationship which brought about the delicate proportions of No 102 *La France*. The Société Alsacienne des Constructions Mechaniques of Belfort in France built the engine and it was the Great Western who got them to anglicise it to suit their requirements by specifying such items as chimney and cab – and, of course, providing the tender, which was pure Great Western, with a 4000-gallon water capacity instead of the more-usual 3500-gallon type widely used with its bigger engines at that time. Delivered in crates, *La France* was bolted together at Swindon and entered service on the GWR in October 1903.

Ever watchful of locomotive performance elsewhere, and aware of the reputation of the Alfred de Glehn-designed compound four-cylinder 'Atlantics' running on the Chemin de Fer Nord in France, Churchward decided to assess their capabilities alongside his own locomotives on the Great Western, so *La France* was purchased and one of his own engines, No 171 – altered from a 4-6-0 to a 4-4-2 (the same as the French engine) and appropriately given the name *Albion* – was chosen for direct comparison. Not entirely satisfied with the result, Churchward then went ahead and designed the Great Western's own four-cylinder engine, the 4-4-2 No 40 (soon to be named *North Star*), which was to be the forerunner of all the Great Western's famous four-cylinder engines ('Stars', 'Castles' and 'Kings') and which, at the time (1906) was to establish beyond doubt that a Great Western locomotive was a force to be reckoned with. In the meantime, two more larger de Glehn compounds had been purchased in 1905, No 103 *President* and No 104 *Alliance*.

Although thoroughly investigated, the compound system (that of using exhaust steam from the high-pressure cylinders to drive larger-diameter low-pressure cylinders) was not adopted by the Great Western (in any case, they had tried it before) as no appreciable increase in power or economy was evident compared with its own engines, which were to remain 'simple' and continued to use steam at high pressure, directly from the boiler, in their cylinders. One wonders, though, if the results would have been quite the same if the valve gears of *La France* and her sister engines had been 'Swindonised' and the valves set in the way their own engines' valves were set. And did those old diehard Great Western enginemen *really* juggle with the two sets of valve gear controls to give the 'fine tuning' demanded by the French engines' separate pairs of cylinders?

Quite a number of features, however, were adopted by the Great Western from the de Glehn compounds, most important of all being that of 'divided drive', which meant that the outside cylinders drove the second pair of driving wheels, and the inside cylinders drove the leading pair by way of a cranked axle, this arrangement being used on *North Star* and all the subsequent four-cylinder engines built at Swindon. One little detail not adopted, though, was the arrangement of the overflow pipes from the injectors situated below the cab. On Great Western engines, these overflow pipes led outwards under the bottom footsteps as a rule, so that they could be viewed from the footplate by leaning out of the cab. If there was no tell-tale water discharge the injector had 'picked up' and was working properly. Not so the French engines. With that little touch – aimed, no doubt, at creature comfort – their overflows came up into the cab where, open ended, they discharged into funnels draining away downwards so that they could be observed by crews not having to 'stick their heads out of the windows' – a nice Gallic gesture but one which must surely have filled the cab with steam whenever an injector faultered.

La France ran for a while painted black with red-and-white lining, as seen here. Later, she was painted in the green and Indian red livery of the day and ran with either type of tender – the original or one of the 3500-gallon pattern. Later still, she was reboiled 'Swindon fashion' with a taper boiler and adorned with a copper-capped chimney and brass safety valve bonnet. Having made a lasting impression on British steam locomotive history, *La France* continued to grace the Great Western Railway until 1927 when she was retired from service.

In her heyday, though, clad entirely in black, *La France* must surely have cut an elegant dash out on the main line, with her slender appearance and dainty, almost capricious, valve gear dancing round – a belle dame indeed, whether the lady was on trial or not.

GWR No 102 *La France*

WITH the introduction of his first four-cylinder engine No 40 *North Star* in 1906 as a result of the trials with the French compounds, Churchward had produced a locomotive design far in advance of anything running on British rails at that time – his masterpiece – and one which was to set the pace for four-cylinder locomotive practice (on other railways as well as the Great Western) for many years to come, starting with the 'Star' class.

Never shy in telling the world at large about the virtues of the West Country and its holiday resorts, the Great Western had reaped a harvest of increased passenger traffic year by year, so the 'Stars', as a class, came into being as the powerful mainstay for these ever-heavier West of England trains. Following the pattern of *North Star*, the new 'Stars' were built from 1907 onwards, the first batch of ten starting with 4001 *Dog Star* and going on to perpetuate the names borne by those very first illustrious pioneer Great Western engines, *Evening Star*, *Lode Star*, etc. These were now built as 4-6-0s, the 4-4-2 arrangement of *North Star* being brought into line with the rest of the 'Stars' as a 4-6-0 two years later, and given the new number 4000 in 1912.

In less than ten years, Great Western engines under Churchward had gone from an elegant appearance through a phase of gaunt – even ungainly – design in the interest of increased efficiency, to arrive at the balanced and functional features of the 'Stars', whose appearance, it may be said, was as well-proportioned a design as any ever applied to a 4-6-0 engine. Apart from the difference (initially) in wheel arrangement and a 'cleaned up' appearance by having rounded drop-down sections under the smokebox front and cab at the extremities of their running plates ($2\frac{1}{2}$ in. lower than on *North Star*), the major difference between the prototype and the production 'Stars' was in the valve gear, Walschaert's gear now replacing the unique 'scissors' arrangement used in *North Star*, whereby the travel of each of the two inside cylinder valves was derived from the movement of its neighbouring crosshead. Both systems had rocker arms linking inside valve movement to the outside cylinder valves.

Prince of Wales was one of the later 'Stars' to be built in 1913, by which time the first forty engines of the class had appeared, these being turned out ten at a time, *Prince of Wales* heading a list of five engines named after Princes to be accompanied by fifteen 'Princesses' the following year.

By the time *Prince of Wales* was built, superheating (the system whereby steam is re-heated on its way to the cylinders to become dry steam at a much higher temperature than the wet steam taken directly from the boiler) had become established in the 'Stars', the increased efficiency it produced allowing the cylinders of *Prince of Wales* and subsequent 'Stars' to be opened up from $14\frac{1}{4}$ to 15 in. bore and gain an increase in tractive effort from 25,100 to 27,800 lb – a substantial increase in power from what was still the same basic locomotive and one which was to keep the 'Stars' at the top of the power league for passenger engines for some years to come.

Prince of Wales was chosen to illustrate the 'Star' class as it shows the outline shape to which the design in general settled down. Not quite the ultimate look of the 'Stars', though, as the class was rounded off by twelve more engines following a break in building during the First World War. These were the 'Abbey' series, built in 1922/3 and which, in the aftermath of wartime economies, came out as austere looking engines with plain chimneys and no brass beading bordering their splashers, all other 'Stars', including *Prince of Wales*, having already had their splasher embellishments removed, never to get them back.

As with any class of railway engine, alteration in detail to the 'Stars' was manifold over the years. Bogie brakes were removed, wooden roofs gave way to steel and the small balance weights on the leading driving wheels disappeared when balanced cranks were introduced with the 'Abbeys'. In later years, many 'Stars' sported outside steam pipes, some akin to those of a 'Castle' and others as a knuckle turning inward to feed into the backs of the outside cylinders. Larger 4000-gallon tenders appeared with many of the 'Stars' in later years instead of the 3500-gallon pattern seen here.

Despite the introduction of the more powerful 'Castles', the seventy-three 'Stars' continued to take their share of the heavy express work load and, discounting the fifteen rebuilt as 'Castles', forty-seven of the class still remained in service to pass into British Railways' ownership in 1948 at the demise of the Great Western Railway. Happily, 4003 *Lode Star*, an example of one of the original batch of 'Stars', resides today at its birthplace as a static exhibit in the Great Western Railway Museum at Swindon, sole survivor of the class of engine which provided the world with such a great step forward in locomotive technology, and one worthy of the name of 'Star'.

GWR No 4041 *Prince of Wales*

*T*HE *Great Bear* – the enigma. Why, in 1908, this engine came to be built has been much discussed by railway historians and will, it seems, remain a riddle in the history of the Great Western Railway. Had the directors (and its originator, Churchward) some particular purpose in mind for such a large engine or was it simply meant as a prestigious showpiece for reasons of publicity? It was, indeed, a massive locomotive, and the only 'Pacific'-type (4-6-2) tender engine to run on British main-line rails for a period of fourteen years. The GWR certainly made use of it to the utmost advantage in its publicity material – but why so big?

The power *The Great Bear* produced was by no means proportional to its extra size as its tractive effort was only fractionally higher than that of the 'Stars' in service at the time when it was built, many of whose parts (wheels, cylinders, bogie, etc.) were embodied in *The Great Bear*'s design. Indeed, later 'Stars' with their cylinders opened up to 15-in. diameter (see *Prince of Wales*) were equal to *The Great Bear* – and with much smaller boilers to meet adequately the demand for steam made on them in traffic at that time.

So why such a large boiler? And one requiring a large firebox, too, of a wide pattern, in order to gain sufficient grate area to complement the proportions of such a boiler – in direct contrast to the narrow fireboxes which had by now become established policy for Great Western boilers, and within whose confines Welsh steam coal was burning so efficiently to contribute to their outstanding success. Perhaps, though, Churchward was looking even further ahead than he is normally given credit for doing and *The Great Bear* was meant as his precursor for even mightier locomotives, and was simply using existing parts of 'Stars' as an expedient for economy's sake in this, his 'test piece'. Or was he out to prove (and this is the theory to which I subscribe) that voluminous boilers, favoured by many locomotive engineers of the day, were not a necessary requirement for producing the large amounts of steam needed for sustained heavy haulage? If this were the case, then Churchward had killed

two birds with one stone – by proving his theory correct and, at the same time, giving the Great Western the largest engine in the country with which to do it. One can almost hear him saying, 'There – I told you so.'

The tender of *The Great Bear*, carried on two shortened versions of locomotive bogies, seemed never quite to complement the massive bulk of the engine, despite its being Swindon's largest tender so far, and because of the overall length of the engine and tender (71 ft $2\frac{1}{4}$ in., buffer head to buffer head) combined with a heaviest-yet axle loading of 20 tons, *The Great Bear* was soon scheduled to show its paces mainly between London (Paddington) and Bristol – an easy road for the most part, and one not likely to stretch the mighty engine to its full capacity. Beset by teething troubles at first (the steps ahead of the cylinders were removed almost immediately as they fouled the platform at Paddington), *The Great Bear* settled down in time and seems to have behaved itself reasonably well, though it does appear to have been bedevilled with troublesome trailing wheels whose bearings, being directly under the firebox, were rather prone to collecting ash. Its usefulness on the Bristol run lasted until 1924 when a complete boiler replacement became due, and by which time the new 'Castles' were being exposed to the full limelight of publicity, so the one-time pride of the Great Western Railway was dismantled and many of its parts (including its '111' number plate) used in the construction of a new 'Castle' class engine, *Viscount Churchill*.

Here, *The Great Bear* is shown in its very early condition, still with its forward steps intact and with elongated, wooden cab roof (later shortened, it is said, because long fire-irons got themselves stuck under its lip when in use) and before being fitted with top-feed to the boiler. Despite being considered by many as being not quite as handsome an engine as one might expect from the Great Western, *The Great Bear* was nevertheless an endearing locomotive, and one which created an immense amount of interest – whatever the reason for its being built.

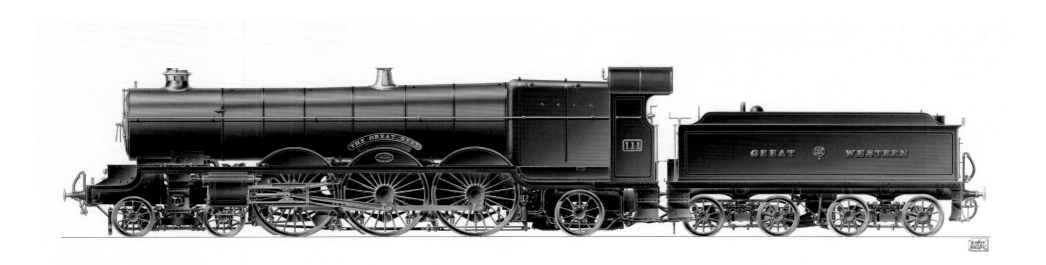

GWR No 111 *The Great Bear*

GER No 1900 CLAUD HAMILTON

RESPLENDENT in deep royal blue trimmed with red, and with an abundance of polished brass embellishing its rich splendour, the GER's *Claud Hamilton* was no sooner completed, when it was shipped to France to gain the distinction of being awarded the Grand Prix at the Franco-British Exposition of 1900, held in Paris – a mark of excellence which brought with it a gold medal for its originator, James Holden, and silver and bronze medals for members of his staff responsible for the engine's design.

Numbered 1900 to coincide with the year of its birth (it was actually rolled out of Stratford works on St Patrick's Day, 17 March 1900), subsequent engines of the class had numbers which ran, not forwards, but in a backward sequence – 1899, 1898 etc., and finishing with No 1780, a total of 121 engines – and although the later ones built differed quite considerably in detail from the original batch of eleven produced in that first year, they were all based on the *Claud Hamilton* design and became known fondly as the 'Clauds'. All of them carried the ornate cast-brass company crest (picked out in colour) on their splashers but only the first one carried a name, and this in honour of the chairman of the Great Eastern Railway at that time, Lord Claud Hamilton.

In small letters under the nameplate of *Claud Hamilton* were engraved the words HOLDEN'S PATENT, this referring, not to the engine itself, but to the equipment Holden had devised and patented for firing his locomotive with oil fuel. This method came about some years before, when the Great Eastern Railway had been brought to task for discharging waste oil into a local river at Stratford on the eastern outskirts of London, this oil being a by-product of the plant which manufactured the oil-gas used for lighting carriages. With the Law threatening and a surplus of oil on his hands, Holden hit on the idea of burning the stuff in his engines, and altogether some sixty engines, including *Claud Hamilton*, were equipped with his device to burn the oil which otherwise would have gone to waste.

The oil-burner principle was simple enough. Using coal, a fire was first lit in the engine, and when sufficient steam was raised to inject the oil into the firebox through nozzles, the fire was covered all over with a layer of broken-up firebrick pieces, and on to this incandescent firebed the oil was sprayed, where it ignited. This method, beside using up the oil (as well as saving on coal bills), had the added advantage that the engines could be readily converted back to coal-burners – which, in fact, they were when coal-gas from town supplies became available to replace oil-gas in Great Eastern carriages, coupled with the cost of the oil bought to supplement the home-produced supply rising sharply. In time, *Claud Hamilton*'s unusual rounded-top tender, holding oil, water and a supply of coal, was replaced with an equally short high-sided conventional coal-carrying tender, but a handful of 'Clauds' ran for years with their original tenders adapted for carrying coal.

Another interesting, and novel, feature of *Claud Hamilton* was the way it made use of compressed air. In common with other Great Eastern engines it was equipped with Westinghouse air brakes, the brass-bound pump on the side between its splashers compressing the air into a reservoir which was the large hollow box cross-member stretching across the frames under the cab. Not only the brakes, though, but the reversing gear and the water pick-up scoop under the tender, too, were compressed-air operated, with provision for both to be hand-operated if required.

In time, the 'Clauds' found their way all over the Great Eastern system, and although the long easy stride of their 7-ft driving wheels was ideally suited to the broad flatlands of East Anglia, it was on the strictly timed expresses on the somewhat more tortuous road out of London (Liverpool Street) to Colchester and beyond where they were really put to the test, handling such crack trains as the heavy 'Norfolk Coast Express' with flying colours.

During its lifetime, *Claud Hamilton* underwent considerable changes, losing its splendid blue livery for one of apple green soon after the Great Eastern had become part of the newly formed London & North Eastern Railway in 1923. A new Belpaire boiler came its way in 1925 and in the 1930s the locomotive was completely rebuilt, losing nearly all its distinctive features in the process. Away went the ornate slotted valences over the driving wheels, the brass-topped chimney and all the brass trim, including the original engraved nameplate, which was replaced by one of a cast brass pattern, arched above the leading splasher. Thus rebuilt, complete with Gresley boiler, chimney and cab, *Claud Hamilton* survived until 1947 after which, as a fitting tribute to a noble class of locomotives, the nameplate was carried forward into the BR era on engine No 62546, one of *Claud Hamilton*'s original classmates, No 1855 of Great Eastern days, and a 'Claud' which was to remain in service until 1956.

GER No 1900 *Claud Hamilton*

LBSCRly No 326 BESSBOROUGH

THERE were two of these big 'J' class 'Pacific' tank engines on the London Brighton & South Coast Railway: the first, 325 *Abergavenny*, being built at their Brighton works in 1910 under the supervision of Locomotive Superintendent D. Earle Marsh, and although he must be given credit for much of the substance of the second engine, 326 *Bessborough*, it was his successor, L. Billington, who put the finishing touches to the design in the months when he was standing in for Marsh during the illness which led to his retirement in December 1911. *Bessborough* was completed three months later in March 1912, and although built to the general – and handsome – proportions of its sister engine, it did differ quite a bit in detail. Most noticeable was the Walschaert's valve gear, carried on the outside of the engine and working through a rocker-arm linkage to the valves which, following the pattern of *Abergavenny*'s inside Stephenson's valve gear configuration, were mounted behind the cylinders, between the frames. Of the two engines, 'Bess' was considered as having the edge over its sister, being fractionally faster in traffic with a somewhat brisker acceleration.

Named in honour of Lord Bessborough, chairman of the LBSCR, *Bessborough* ran for a while in works' grey paint before being decked out in the company's livery of the day, a rich deep brown (umber) generously lined out in black edged each side with a thin yellow line, the lining on the tanks and bunker sides coinciding with the rows of bolt heads holding their cladding plates in place. This livery complemented the dark brown and off-white of the sumptuous Pullman cars making up the luxury express trains which *Bessborough* and *Abergavenny* had been specifically designed to handle between London (Victoria) and Brighton.

Bessborough, whose water capacity was the lesser of the two engines, was fitted with a condensing system (along with quite a number of LBSCR tank engines) not unlike that which had been in use on the GWR, whereby steam was taken back into the vertical columns (one each side) seen standing proud at the front end of the tank top, where it condensed, any residue of steam being ventilated through the pipes leading up and turning over the front edge of the clerestory-roofed cab. These pipes were carried only a short way originally, but were later lengthened along the roof, suggesting that steam from them tended to obscure the vision of the crew. Problems, too, seem to have been encountered with the fouling of platforms by the buffer beams, due to the length of the engine, as these had cut-outs in the shape of an inverted J removed from their bottom corners soon after she had entered service. A detail worthy of note on 'Bess' is the hinged flap on the tank side, this giving access to a box section let into the bottom of the tank (accounting for the reduced water capacity) which accommodated part of the Walschaert's valve gear.

As happened to so many of those handsome locomotives turned out in the early years of the century, characteristic features, so much a part of the original – and pleasing – design, disappeared as the years went by and engines no longer fulfilled the role for which they had been designed. Among the first things to go from *Bessborough* was its name, this being blotted out when the engine was painted green in 1925. It was replaced by the word 'Southern' as the old LBSCR had now become part of the recently formed Southern Railway, in whose service *Bessborough* became plain 2326. The ornate clerestory roof went ten years later in 1935, to be replaced by one of a lower elliptical pattern, blending with the cabsides, and at the same time the delicately proportioned chimney and dome gave way to those of a more squat nature. These modifications gave the engine a much wider running range over the Southern Railway, on lines whose loading gauge was less generous than that of the LBSCR. Progress, in the form of electrification, had, by then, overtaken both 'J' class engines, firstly on their Brighton run, and now on the Eastbourne line where both had gone to take charge of such trains as the 'Sunny South Express' and other through trains between the Midlands and the South Coast.

Thus cut down to size, both engines remained at Eastbourne, then spent much of their wartime years in and around London before moving out to work services around Tunbridge Wells. Arriving at Tunbridge Wells West shed in 1946 after overhaul at Eastliegh, and still wearing austere wartime black, an understanding shedmaster persuaded the powers-that-be to paint them green again, so back they went to Eastliegh. Both engines remained Southern green to the end of their days, 'Bess' being withdrawn from service as BR 32326 in June 1951 with nearly a million miles 'on the clock', to be cut up on 6 July. Six weeks later, *Abergavenny* suffered the same fate.

BESSBOROUGH

326

LBSC Rly No 326 *Bessborough*

LNER No 4472 FLYING SCOTSMAN

'HE could have had ours!' Churchward is reputed as having said when he learned that H. N. Gresley (later, Sir Nigel) had introduced his new three-cylinder 4-6-2 No 1470 *Great Northern* on the Great Northern Railway in 1922. He was, of course, referring to *The Great Bear*, the Great Western's solo performer which had had the stage to itself for the previous fourteen years as the country's only main-line 'Pacific'-type tender engine and which was now about to be upstaged by this new breed of handsome Gresley 'Pacifics'.

Following the pattern he had established in his recently introduced large three-cylinder 2-6-0 No 1000 Gresley had, with his new engine, made use of the system whereby the outside cylinder valve movements were transmitted to the inside cylinder valve through a two-to-one ratio linkage – not a new idea but thereafter known as the Gresley conjugated gear.

The following year, 1923, saw the formation of the London & North Eastern Railway with the Great Northern Railway as a major constituent and Gresley in overall charge of locomotive development as Chief Mechanical Engineer. In February of the same year, the third in line of the new 'Pacifics' appeared, this being No 1472 – a locomotive destined to become one of the best-known and best-loved of all railway engines, being re-numbered 4472 and given the name *Flying Scotsman* a year later when it was chosen to represent the LNER at the British Empire Exhibition at Wembley.

Embellished with a certain amount of brass trim (not carried by sister engines), noticeably around its splashers and boiler washout fittings, and with polished wheel rims and axle ends (again, not standard), *Flying Scotsman* stood at the exhibition in close proximity to the Great Western's 4073 *Caerphilly Castle*, and where the GWR, ever tractive effort conscious, proclaimed their less-massive exhibit as being the most powerful passenger engine in Britain. Friendly rivalry led to a serious exchange of locomotives (see 4079 *Pendennis Castle*), resulting in the Great Western eclipsing the LNER in locomotive performance and Gresley, swallowing his pride, modifying the valve gear of his 'Pacifics' to give longer valve travel more akin to GWR practice – and in doing so, producing a class of heavy-hauling locomotives *par excellence*.

With the introduction of the Flying Scotsman express on 1 May 1928, running in both directions between London (King's Cross) and Edinburgh (Waverley) as the longest non-stop scheduled run in the world, it was fitting that *Flying Scotsman*, still carrying its brass embellishments, was one of the locomotives chosen to head the train, and this is the condition in which *Flying Scotsman* is shown here, complete with one of the newly introduced corridor tenders which made the non-stop schedule possible by allowing crews to be changed whilst on the move.

The zenith of *Flying Scotsman*'s career, though, came a few years later, when on 30 November 1934 (in anticipation of the high-speed 'flyers' to come) it took a special test train from King's Cross to Leeds and back. In charge was driver William Sparshatt, a man renowned for getting the very best performance out of a locomotive, and on the return run he made sure that *Flying Scotsman* was to become the first British steam locomotive to be *officially* recorded at 100 mph, to nudge the magic figure for a few seconds only as he brought the train dashing down Stoke bank between Grantham and Peterborough – but long enough to gain for *Flying Scotsman* a niche in railway history.

During 1927, a new brand of 'Pacific' had appeared – the so-called 'Super-Pacific' (classified 'A3') when two of the class, 4480 *Enterprise* and 2544 *Lemberg* were modified and, along with other alterations, fitted with 220 psi boilers to replace the 180 psi boilers carried by *Flying Scotsman* and its sisters (classified 'A1'). Although similar in appearance, these extra-powerful A3 'Pacifics' were readily identified (certainly by train spotters) by their having a square protruding 'patch' high up on each side of the smokebox (it housed the ends of the new extra-width superheater header) and although most of its sister engines became converted from A1 to A3, *Flying Scotsman* ran throughout its career as 4472 minus these patches, as it was not until 1946 that it was reboilered and became an A3, by which time it was renumbered 103.

Retiring in 1963 with well over two million miles to its credit, *Flying Scotsman* is now owned by the Hon. William McAlpine, leading an extremely active life in preservation today, clad in its original apple-green livery and still wearing its old number, 4472. Apart from now having a banjo-shaped dome in place of the original round one shown here, and with smaller cabside cut-outs, *Flying Scotsman* looks much as it did in those halcyon days when steam was the master of the East Coast Route.

LNER No 4472 *Flying Scotsman*

GWR No 4079 PENDENNIS CASTLE

and the 'Castles'

TAKE a 'Star', increase the cylinder bores to 16 in. diameter, put on a bigger boiler and lengthen the frames to accommodate a new and larger cab, and there you have – a 'Castle'. Coming immediately on the heels of the 'Abbey' series and taking up the number where the 'Stars' left off, 4073 *Caerphilly Castle*, first of the new 'Castle' class, made its debut in late August 1923 under the hand of Mr C. B. Collett who had now come to power as Chief Mechanical Engineer at Swindon.

Following in the master's footsteps, Collett had taken Churchward's 'Star' design a step forward and by doing so, in terms of its 31,625-lb tractive effort, had produced once again for the Great Western, Britain's most powerful passenger locomotive – a fact which their publicity department were not slow in proclaiming, getting their message across in a new 'Shilling Book for Boys of All Ages' entitled 'Caerphilly Castle' almost immediately. Its publication coincided with the new locomotive's appearance at the British Empire Exhibition in 1924 and in which, although mentioning the proximity of George Stephenson's *Locomotion*, no word was made of the LNER's *Flying Scotsman*, whose presence there with *Caerphilly Castle* was to spark off such rivalry between the two railways.

Most noticeable from established GWR practice on the new 'Castles' was the new pattern of cab with its brass-bound side windows and extended roof. Side-windowed cabs had been used before on the Great Western (4-4-0s 3292 *Badminton* and 3297 *Earl Cawdor* had both carried them) but never perpetuated, and one is left wondering how much they owed their revival to the influence of the two larger French compounds, 103 *President* and 104 *Alliance*, whose cabs had been side-windowed, too, and not entirely unlike those of the new 'Castles'.

If any part of the 'Castle's' overall design, initially, fell short of expectations it was in the shape of its tender which was still of the 3500-gallon pattern at first – the same one which had looked so 'at home' on a 'Star' but which now did little to enhance the heavier look of a 'Castle'. It was not until the third batch of ten 'Castles' were built in 1925 that the matter was rectified when 5000 *Launceston Castle* appeared and became the first Great Western engine to receive one of the new high-sided 4000-gallon tenders seen opposite. In time, all 'Castles' got them, but not before 4079 *Pendennis Castle* had gone over to the

LNER as a contestant in the power challenge of 1925 initiated at Wembley, armed with its smaller tender to look even more diminutive alongside the rival's 'Pacifics' it was to challenge. Honours went to 4079 *Pendennis Castle*, both in power and economy of fuel – as happened, too, to 5000 *Launceston Castle* when its turn came in 1926 to venture forth on LMS metals for similar assessment.

With their ample power, brought about by a combination of balanced machinery and well-planned four-cylinder design supported by the economical and free-steaming quality of their boilers, there can be no doubt that the 'Castles' were among the most successful classes of locomotives ever to be built. As standard main-line express engines of the GWR they were equally suited to the long-haul heavy passenger trains to and from the West Country, South Wales or the West Midlands, as they were when making the flying dash down the Vale of the White horse and along the Thames Valley on the Cheltenham Flyer between Swindon and Paddington, to justify with apparent ease the Great Western's claim that this was the fastest train in the world during the early 1930s.

So successful were the 'Castles' that they continued to be built over a period of twenty-seven years, the last one, 7037, being appropriately named *Swindon* by Her Majesty The Queen (then The Princess Elizabeth) in 1950. Appropriately, too, the last 'Castle' to be built under the auspices of the GWR in 1947 before nationalisation was 7007 *Great Western* – a revival of the name of the very first engine to be built in its entirety at Swindon in 1846.

Altogether, 155 'Castles' were built, added to which were sixteen rebuilds – fifteen converted from 'Stars' (including 4000 *North Star*, still with its running plates $2\frac{1}{4}$ in. higher than the rest but now fitted with Walschaert's valve gear) and the reconstruction of *The Great Bear*, No 111 *Viscount Churchill*.

Eight 'Castles' are in preservation today and include the original engine 4073 *Caerphilly Castle* of 1923, immaculately presented as a static display at the Science Museum in London, whilst more active are 5051 *Dryslwyn Castle* (at one time *Earl Bathurst*) of the mid-1930s and 7029 *Clun Castle* of the 1950s British Railways period. Much travelled and still very active indeed is 4079 *Pendennis Castle*, which nowadays has its home in the care of Hamersley Iron Ore Pty Ltd at Pilbara in Western Australia.

GWR No 4079 *Pendennis Castle*

GWR No 5069 ISAMBARD KINGDOM BRUNEL

and later 'Castles'

NO excuses are offered for including two illustrations of 'Castles' within these pages. Besides wishing to show differences between 'Castles' over a fifteen-year period, *Isambard Kingdom Brunel* was chosen originally as a personal tribute to the great man during the 1985 150th anniversary of the Great Western Railway.

A major change had taken place at the front end of the main frames some time before *Isambard Kingdom Brunel* was built in 1938, at the point where they had originally been stepped or 'joggled' inwards in order to accommodate the side-to-side movement of the leading bogie wheels. Cracking had occurred in the weakness produced by this joggling, so from 4093 *Dunster Castle* onwards the frames had no joggle but a dished section was pressed into the frames to give clearance. The bogie itself was now of a somewhat heavier pattern generally and, although not visible in the side view, had additional half-diagonal bracing bars across the front of its frame. The inside cylinder block, instead of having a fluted section removed from its top edges, was now of a plain rectangular pattern (as from 5013 *Abergavenny Castle*) with deep valve rod end covers now reaching down to the footplating, as on a 'King'.

The handsomely proportioned copper-capped chimney had now taken on a slightly more squat appearance, being 3 in. shorter from 5043 *Barbury Castle* (later *Earl of Mount Edgcumbe*) onwards, and at the same time the lamp bracket, formerly mounted on top of the smokebox, was moved to the more accessible position on the smokebox door.

Beneath the cylinders, pipes from the drain cocks had been lengthened and gathered together at a bracket at the front end. On the motion side, the crosshead, instead of being made up from separate parts and bolted together, was now a one-piece casting, without the outside oilbox for its connecting rod pin, a raised section at the little end of the connecting rod where it enters the crosshead now taking care of lubrication by containing an oil box. Oiling arrangements for all the moving parts around the cylinders of a 'Castle' seem constantly to have been undergoing change, those shown here having been drawn in from the works' photograph of *Isambard Kingdom Brunel*.

At the other end of the connecting rod, and again from 5013 onwards, the knuckle joint of the coupling rod was now no longer hidden behind the connecting rod, but had taken up position to the rear of the crankpin to make it more accessible. Wheels are of a different pattern, too, no longer having their crankpins and bosses in line with a spoke, but now central to a pair of spokes with a solid web cast between them. The wheels, although individually sprung on early 'Castles', had compensating beams working between their springs, but now have individual 'J' hangers at each end of the springs, suspended by long bolts secured through steel cups containing hard rubber cushioning pads.

From 4093 onwards, the sanding gear originally fitted to the centre pair of driving wheels was removed and replaced by back-sanding pipes to the rear wheels, their sand boxes housed high up on the frames out of sight behind the lower part of the cabside sheets, but from 5043 onwards the boxes appeared lower down in the more accessible position seen on *Isambard Kingdom Brunel*. Water pipes leading from injectors up to the top feed became deflected, too, now taking a path over the tops of the wheel splashers behind the nameplates instead of coming out from behind the frames directly under the covers carrying them upwards to embrace the boiler, as on earlier 'Castles'.

The long fire-iron box with its opening inside the cab, seen with its top in line with the tops of the rear splashers, first made its appearance on 5013, while on the cab itself a double row of horizontal rivets just below the window located a reinforcing strip inside the side sheet, being introduced from 5043 onwards. Whistle shrouds to prevent steam obscuring forward-facing windows were fitted to most Great Western engines over a period of years, and by 1936 all the 'Castles' then built had them.

Double chimneys, re-shaped steam pipes with easier curves and mechanical lubricators on the right-hand side running plate ahead of the steam pipe came later to many 'Castles' in British Railways ownership, as did flush-sided tenders after they had been introduced by the GWR in 1945.

Isambard Kingdom Brunel had but one alteration made before it reached the state in which it is shown here, this being concerned with the nameplate which originally had been in the shape of a flat-arched design – incongruous in that it did not follow the radius of the splasher on which it was mounted. Whether or not it ever ran thus adorned, or whether it was realised that the volume of letters could be carried – just – on a traditionally shaped nameplate before it ventured forth from Swindon, seems not to have been recorded.

GWR No 5069 *Isambard Kingdom Brunel*

REFERENCE has already been made to how the 'Super-Castles' – the 'Kings' – came into being. Most publicity-conscious of all the railway companies, the Great Western had, in the past, set great store in proclaiming the superior power of its locomotives in terms of tractive effort (which, it may be argued, is not the only criterion by which a locomotive's abilities should be judged) so, with the Southern Railway's 'Lord Nelsons' stealing the limelight, the GWR felt something ought to be done about it. The resulting 'King' was a locomotive displaying all the power and glory Swindon could muster, with a tractive effort of 40,300 lb, which was considered way out in front of anything its rivals were likely to achieve for years to come.

According to legend (one still perpetuated when I served on the GWR) it was never the intention to call the new locomotives 'Kings' at all. With 'Abbeys' and 'Castles' firmly established in the power chain, it was logical that the next link was to have been a 'Cathedral' and this, in all probability, it would have been had not the Great Western been invited to send a representative locomotive to the centenary celebration of the Baltimore & Ohio Railroad in the USA in 1927. All haste was made to complete the new 'Super-Castle' in time, and as ambassador of Great Britain as well as the Great Western, it was considered appropriate to honour the locomotive with the name of the reigning monarch, King George V – a master-stroke of publicity if ever there was, and one used to even greater advantage when *King George V* came home, heaped with honour and wearing a brass bell and cabside medallions to commemorate the visit.

The 'Lord Nelsons' were not the only reason for the 'Kings' being built as, for some time, the Great Western had been exploring the possibility of an even more powerful passenger engine than a 'Castle', having long since discounted any development along the lines of *The Great Bear*. Although regarded initially as 'Super-Castles', the 'Kings' were not strictly straightforward developments of 'Castles' in the same way that they, in their turn, had been created directly from the 'Stars'; nevertheless, there was no denying the strong family likeness. The 'Kings' conformed to that same four-cylinder divided-drive pattern established so successfully with the 'Stars', with slightly smaller driving wheels of 6 ft 6 in. diameter, cylinders increased to 28 in. stroke by $16\frac{1}{4}$ in. bore (later ones were 16 in.), and the whole engine stretched to accommodate the new and much larger boiler required for such a powerful engine – one working at an unprecedented high pressure for a GWR locomotive of 250 psi, and shared only by the 'Royal Scots' of the LMS at the time. This meant a longer wheelbase and a proportional increase in bogie-wheel centres, the bogie itself being of particular interest. Plate-framed along the lines of the bogie of de Glehn compound *La France*, it was shaped to embrace its wheels on the inside at the front end and carry them outside at the rear, in order to give clearance to inside and outside cylinders respectively. This unique bogie did much to enhance the solid, thrusting front-end appearance of a 'King' – as did the copper-capped chimney, squat in shape in order to keep overall height within the loading gauge.

The new bogie was not without its teething troubles and newly built 6003 *King George IV* was to highlight its shortcomings when its own bogie derailed at speed as it headed the Cornish Riviera Express. Additional coil springs to both bogie axles solved the problem (those fitted to the leading wheels being seen alongside the leading axlebox) and this modification, along with attention to the springing of the rear coupled-wheels went a long way to placing the 'Kings' among the most sure-footed and easy-riding of Great Western engines.

There were thirty of them altogether, these majestic 'Kings', built between 1927 and 1930, capable of working the heaviest West of England trains (including the 'Limited', the Cornish Riviera) over the South Devon banks single-handed and without the help of a 'double-header', though the weight restrictions imposed by the $22\frac{1}{2}$ tons on each of their pairs of coupled wheels limited the 'Kings' to running as far west as the Plymouth area (Brunel's magnificent bridge spanning the Tamar prevented their going further) and to the West Midlands as far as Wolverhampton, though later they were allowed to run through to South Wales.

Three 'Kings' survive today. At the time of writing, two are still undergoing restoration: 6023 *King Edward II* in the hands of Harvey's of Bristol, and 6024 *King Edward I* nearing completion at Quainton Road, cared for by the 6024 Society. *King George V*, now in BR livery and with double chimney and 'easy curve' steam pipes, still displays its glory on the main line, thanks to Bulmer's Cider of Hereford.

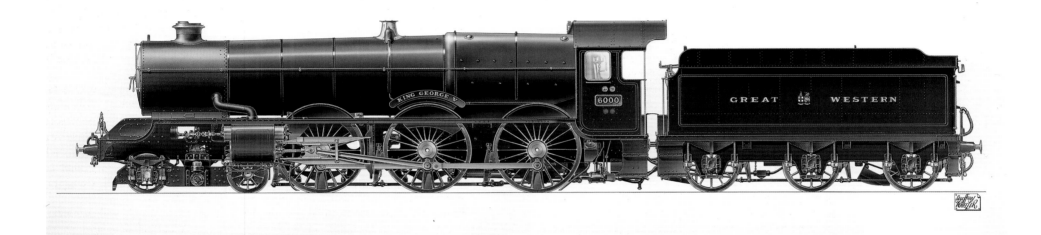

GWR No 6000 *King George V*

GWR No 1450, a 'MARLOW DONKEY'

ONCE upon a time, scores of branch lines wandered all over the countryside to serve a myriad of towns and villages off the beaten track of the main lines, and where diminutive engines pulled (or pushed) little trains which were as much a part of the character of the countryside as the communities they served. Most of these branch lines have now gone for ever, but one which still remains is the ex-Great Western Marlow branch, a single-track line running alongside the Thames on the Buckinghamshire bank between Bourne End and Marlow. Opened in 1873 as the Great Marlow Railway and originally broad gauge, it had, at some time during its history, acquired for its locomotives the local name of 'Marlow Donkey', no matter which one plied its $2\frac{3}{4}$-mile length, it seems.

No 1450, this particular 'Marlow Donkey', worked on the Marlow branch around 1946, being one of quite a number of these little 0-4-2 tank engines of the same class to do so. Introduced by Charles Collett in 1932 to replace the ageing '517' class tanks to which they bore a strong resemblance, these engines were of a slightly less weighty nature than the majority of the smaller Great Western tank engines to give them availability on all branch lines, and were suitably equipped for working auto-trailers – open saloon coaches, known to many a Buckinghamshire schoolboy as 'agony wagons' due, no doubt, to their hard horse-hair stuffed seats.

Auto-trailers were either pulled or pushed by the engine, or arranged with the engine in the middle of a pair (or more) in a pull-and-push situation – again with a local reference, that of a 'put-and-take'. Working in the direction with an auto-trailer leading, the driver would leave his cab and take up station in the front end of the coach to control the train, his fireman remaining on the footplate of the engine. A regulator lever in the coach compartment was connected back to the engine's regulator by a mechanical linkage, the couplings for this on No 1450 being visible just below both buffer beams, more noticeable at the rear end alongside the screw coupling.

Altogether, seventy-five of these auto-train 0-4-2 tank engines were built, numbered at first 4800–4874, to be supplemented by a further batch of twenty engines built without auto-train equipment for light passenger and goods work on branch lines and numbered 5800–5819. With the introduction of oil-firing directly after the Second World War, some of the big 'Twenty-eight'

2-8-0 mineral engines were converted to burn oil, taking on new numbers as the 48xx series, so the class of auto-fitted 0-4-2 tanks became the 'Fourteens' and were given new numbers, 1400–1474.

There were two noteworthy differences between engines of the class. Boilers on many of them had top-feed fitted, supplied by pipes running along each side of the boiler from the tank tops, these turning upward to feed clack valves contained in a tall casing mounted crosswise between the chimney and dome. This arrangement was noticeably absent on No 1450 when it fulfilled the role of 'Marlow Donkey', having its feed pipes and clack valves on the backhead of its boiler, inside the cab. The other difference lay in the positioning of the toolboxes alongside the wheel splashers, the one on the far side (in the illustration) being moved forward on many engines in order to obscure the spare head lamps carried on brackets on that side, thus avoiding any confusion which might arise from a white-painted spare lamp being mistaken as being part of the true head-lamp arrangement indicating the classification of the train. Not all toolboxes were repositioned; some engines had one or both moved forward and others continued to run with both toolboxes in line with the splashers. The other two boxes seen on the engine – those to the rear of the trailing wheels – housed batteries for the ATC (Automatic Train Control) equipment, fitted to all 14xx engines and a number of the 58xx series.

Chimneys were plain with no copper cap; not all brass safety valve bonnets were polished either, as officially the whole of the upperworks – excluding smokebox and roof – were painted middle chrome green in Great Western days. Such was the pride of some enginemen, though (perhaps with time on their hands at the end of some remote branch line), that many brass bonnets were brought back to bare metal and polished. Paint does not adhere too well to brass, as many a model-maker will confirm.

Diminutive they may have been, but these little engines were every bit a part of the Great Western landscape as the mighty 'Castles' and 'Kings', and are regarded with affection, too, as witnessed by the enthusiasm for those four which survive today, including No 1450 which came back to join in the celebration of the centenary of the Marlow branch in 1973 and play the part of the 'Marlow Donkey' again for a day.

GWR No 1450, a 'Marlow Donkey'

LMS No 6201 PRINCESS ELIZABETH

and the 'Lizzies'

ON its formation in 1923 the London, Midland & Scottish Railway, largest of the four main railway companies, had inherited some 6000 locomotives from its constituent companies, not many of which could be considered as being really 'beefy' passenger engines. It was not until the fifty 'Royal Scots' appeared on the scene in 1927 that the situation – particularly on the West Coast route to Scotland out of London (Euston) – started to ease. But not before the Great Western had once again ventured forth on 'foreign' metals, this time with 5000 *Launceston Castle* which had demonstrated its capabilities on the LMS the year before the 'Royal Scots' had been introduced. So impressed were the LMS with its performance that overtures were made to the GWR to try to obtain fifty 'Castles' for their own use, or perhaps the loan of drawings so that they might build their own. The Great Western politely declined, so the LMS went a step further by persuading one of the GWR's top locomotive design engineers, Mr W. A. Stanier (later, Sir William), to join them a few years later as Chief Mechanical Engineer – a very astute move, it so happened, and one which was to carry the LMS into the very forefront of British steam locomotive design.

Stanier took over as Chief Mechanical Engineer on the LMS on 1 January 1932 and within eighteen months his first 'Pacific', 6200 *Princess Royal* (later altered to *The Princess Royal*), had taken the rails, to be joined in November 1933 by 6201 *Princess Elizabeth*, from which engine the 'Princess Royals' became affectionately known as the 'Lizzies'. There was no mistaking the Great Western influence on these two new 'Pacifics' for the principal dimensions of the working parts – cylinder bore and stroke, and driving wheel diameter – were those of a 'King', these combining with a working pressure of 250 psi to give an equal tractive effort of 40,300 lb. The same 'divided drive' was employed, too, though four sets of Walschaert's valve gear were now used (one per cylinder) instead of the 'King's' two sets, along with a bar-framed bogie almost identical to that of a 'Castle' and GWR 4-6-0s.

Nevertheless, Stanier had put his own stamp on the design by making it a 'Pacific' to carry the large boiler with attendant wide firebox to give best results from the grades of coal available on the LMS, to provide sustained steaming for the 500-ton (and more) trains the 'Lizzies' were expected to handle between London and Scotland (as a challenge to the LNER) and which included the long laborious hauls over Shap and Beattock summits on the way. The handsomely proportioned cab with its elongated roof extension was to become a standard feature on Stanier-designed engines.

In 1935 the class was increased to a total of thirteen engines, all but one being similar to – but differing in detail from – the two existing 'Lizzies' and taking their names from princesses and ladies with royal connections. Interestingly, this resulted in two LMS main-line engines running with the same name: 6204 of the 'Princess Royals' and 6004 of the 'Claughton' class, both being *Princess Louise*. The odd one out of the new engines was the experimental turbine-driven No 6203, the 'Turbomotive', which was always included in the 'Princess Royal' class and which became eventually the thirteenth piston-driven engine and named *Princess Anne*, only to be destroyed in the horrendous Harrow disaster of October 1952, eight weeks after conversion.

These new 'Lizzies' had a number of modifications incorporated in their design. Most importantly, their boilers were improved with increased superheating and the provision of longer combustion chambers – that portion of the firebox reaching forward into the water space of the barrel. At one time, a scheme was mooted for these later 'Princess Royals' to be streamlined but this seems to have got no further than the drawing board stage with plans for a scale model of a streamlined 'Lizzie'.

Not quite in its original condition (for a short time it ran with a double chimney) the illustration shows *Princess Elizabeth* as she ran in 1936, by which time a replacement boiler carrying a dome (but still with a short combustion chamber) had been fitted, and the first less impressive tender replaced by one of Stanier's shapely new high-sided pattern of riveted construction (later ones were welded) carrying 10 tons of coal and 4000 gallons of water. In this condition, *Princess Elizabeth* enjoyed what was probably her finest hour when, in November 1936, she took a special train from London to Glasgow and back over a two-day period, reaching 95 mph on both runs and making the return trip of $401\frac{1}{2}$ miles at an average start-to-stop speed of 70 mph.

Two of the class survive today, each a representative of the two batches and both in real life royal sisters – 6201 *Princess Elizabeth* and 6203 *Princess Margaret Rose*.

LMS No 6201 *Princess Elizabeth*

LNER No 4468 MALLARD

and the A4s

THE mid-1930s could well be described as the 'streamlined era', though in most cases it would seem to have been more a question of 'streamlining for streamlining's sake' rather than attempting to achieve any degree of aesthetic excellence. One exception, though, must surely have been the Gresley A4 'Pacifics' of the LNER, which not only achieved a streamlined shape pleasing to the eye, but one which was compatible with the high-speed image of the machine to which it was applied.

In seeking to speed up yet further the trains on the LNER, Gresley had turned his attention to the German diesel 'flyers' of the day, even going so far as investigating the possibility of introducing them on the East Coast Route, before deciding that a British steam locomotive could be designed to do the job just as well, if not better. The result was the A4s.

Mallard was one of the thirty-five superb engines which made up the A4 class. In developing them directly from the A3s, Gresley had polished his 'Pacific' design to near-perfection; any seed sown by the Great Western on the LNER all those years before was about to bear exotic fruit. Nor was it simply a case of applying a streamlined shell to an existing design, for under the air-smoothed casing was a modified boiler, working at 250 psi, with redesigned cylinders of $18\frac{1}{2}$ in. bore (the A3s had been 19 in.) with piston valves opened up from 8 to 9 in. diameter – along with such refinements as improved springing and braking in anticipation of the higher speeds to come. Streamlining was applied inside as well as outside, attention being given to internal steam pipes and passages to give smooth-as-possible steam flow.

Once more, it was a case of the French having some influence on a British locomotive design as Gresley had made a close study of André Chapelon and his work on improved steam flow on the Paris-Orleans Railway, plus the fact that the A4's front end was said to have been inspired by the French Bugatti streamlined petrol-engined railcar's shape, this wedge-shaped front end on the A4s being a relatively simple outline, made up in the main of a 12-ft radius curve centred just above rail level which blended into a 2 ft 6 in. radius curve, then tucking away under the buffer beam to form a nose of 18 in. radius. The resulting contour served a dual purpose by reducing wind resistance at speed, at the same time creating an updraught to lift smoke clear of the crew's vision – a problem which had bedevilled both A1 and A3 'Pacifics'.

First of the new streamlined 'flyers' on the LNER was the Silver Jubilee express, introduced to run between London and Newcastle in the early autumn of 1935. Even before going into service on the train, the first A4, 2509 *Silver Link*, had achieved a maximum of $112\frac{1}{2}$ mph on a trial run, at the same time covering a distance of 25 miles of the journey at an average speed of $107\frac{1}{2}$ mph, a record for a steam locomotive which still stands today. The following year, honours went to A4 2512 *Silver Fox* which nudged the record up to 113 mph. Next, it was *Mallard*'s turn.

Taking its name (as a number of other latecomers to the class had done) from a bird 'strong on the wing', *Mallard* entered service in March 1938 and was the first of four A4s to be fitted from new with the Kylchap double chimney and blastpipe arrangement (again of French origin) to allow steam to be exhausted even more freely to increase performance still further. During spring and early summer of that year, braking tests using A4s were being carried out and on Sunday 3 July 1938 *Mallard* was the locomotive chosen to head the test train out of King's Cross, which this time included a dynamometer car for recording the locomotive's performance. Brake testing was carried out normally on the outward run, and it was only after the turn-round at Grantham to head back towards London that *Mallard* was given its head and allowed to show its paces to the full.

By now, the speed record of 114 mph had gone to the LMS (old rivalries going back to the races of 1888 still lingered between the West Coast and East Coast routes to the north) and somewhere between Corby and Little Bytham, under the charge of Driver Duddington and Fireman Bray, *Mallard* passed this figure, and with speed still rising came sweeping down the bank towards Essendine to reach 125 mph, then hover for a short while at 126 mph – the all-time world speed record for a steam locomotive. The success was not without incident, though, as the ferocious onslaught had caused damage to the engine's inside connecting rod big end – a small price to pay for such a magnificent achievement.

Mallard is presented here as it looked on that, its finest day. At the time of writing, however, *Mallard* is back in action on the main line, less spirited perhaps but looking every bit as if it could do it all again tomorrow, given the chance.

LNER No 4468 *Mallard*

CLOSE on the heels of the second batch of LMS 'Princess Royals' came the 'Princess Coronation' class 'Pacifics', later to become more commonly known as the 'Duchesses'. One can imagine Stanier sharpening his pencil and saying, 'Right! Now let's knock these "Princesses" of ours into shape!' It wasn't quite like that, however. For one thing, Stanier was out of the country while a good deal of the design work was going on (although he had briefed his staff thoroughly and took responsibility for the overall design), and when they did emerge from Crewe works in 1937, starting with 6220 *Coronation*, there was little of the true shape to be seen as they were shrouded in a streamlined casing.

Competition between the East and West Coast routes to Scotland had flared up again. 1937 was the coronation year of King George VI and Queen Elizabeth, and both the LNER and the LMS had launched new fast streamlined trains linking London and Scotland – the Coronation express on the LNER and the Coronation Scot on the LMS. The West Coast route of the LMS to Glasgow was by far the more difficult so their's was the need for an extra powerful express locomotive. This they achieved in the form of the 'Princess Coronation' class, and it wasn't long before *Coronation* itself had wrested the speed record from the LNER at 114 mph.

The first five LMS streamliners were painted blue with silver stripes along their entire length, and the next five to come out in 1938 were crimson and gold. The same year saw the emergence of five more – the *real* 'Duchesses', including 6233 *Duchess of Sutherland* – which were stripped of their streamlined casings to reveal what could be considered the most impressive, most powerful-looking engines ever to run on British rails. These, too, were crimson, but departed from standard LMS big engine livery practice by being lined out in gold, edged on each side with thin vermilion lines.

Soon afterwards, in February 1939, a trial-of-strength took place with one of the new non-streamliners, when 6234 *Duchess of Abercorn* at the head of a twenty-coach train weighing over 600 tons was taken from Crewe to Glasgow and back on a test run over Shap and Beattock summits. A fortnight later, the trial was repeated using the same engine, which by now had been back into Crewe works and fitted with a double chimney and blastpipe arrangement.

The improvement was considerable – so much so that a figure of over 2500 drawbar horsepower was achieved – the highest ever recorded by a British locomotive. This resulted in all the engines eventually being fitted with the double chimney arrangement, *Duchess of Southerland* receiving hers in November 1941.

In terms of tractive effort, the 'Duchesses' were fractionally less powerful than the 'Princess Royals' (40,000 lb compared with 40,300 lb), having larger 6 ft 9 in. diameter driving wheels, but on test *Duchess of Abercorn* had proved what formidable engines they were. The boiler was larger, too, with a cavernous firebox said to stretch a fireman to 'the limits of human endurance'. Some assistance was offered to him by a steam-operated mechanical coal pusher in the tender, to bring coal forward to his firing position. At one time, mechanical automatic stokers were considered for the 'Duchesses' but the idea never materialised.

Despite the impressive outward appearance of the five non-streamlined 'Duchesses', they were the only engines of the class to run as such for the next six years as later additions, even though built throughout most of the wartime years, appeared clothed in streamlined casings, later ones being painted austere plain black. Following the war, however, all were de-frocked. As non-streamlined 'Duchesses' they all received smoke-deflector plates alongside the smokebox, *Duchess of Sutherland* getting hers in September 1946.

A total of thirty-eight engines eventually completed the class, all being LMS-built engines except the last one, 6257 *City of Salford* which was turned out in 1948 under British Railways ownership. This and the penultimate 'Duchess' (6256, aptly named *Sir William A. Stanier, FRS*) differed from the rest of the class by having a modified trailing truck under its firebox, less deep cabside sheets, and modifications made to its reversing gear.

The magnificence of a 'Duchess' in action can still be witnessed on the main line in the shape of 46229 *Duchess of Hamilton*. 46235 *City of Birmingham* resides static in the Birmingham Museum of Science and Industry, while slightly more active is 6233 *Duchess of Sutherland*, nowadays double-chimneyed but without smoke deflectors, at Bressingham Gardens near Diss in Norfolk, often in steam on a length of track it shares with other preserved locomotives.

LMS No 6233 *Duchess of Sutherland*

AS already mentioned, *County of Middlesex* made its debut at Paddington on an August morning in 1945 wearing a double chimney and displaying only its number, 1000. Eventually, a total of thirty similar engines were built. These were the 'Counties', and they differed quite considerably in detail from their long line of two-cylinder predecessors.

F. W. Hawksworth, who was responsible for the design of the 'Counties', had taken over the reins of office as Chief Mechanical Engineer in 1941, at the height of the war, when development of new locomotives on the GWR was at a standstill. Yet he was soon to make his mark (albeit, a minor one) in the implementing of the re-shaping of the front end of the 'Hall' class engines, by running their main frames full-length, right up to the front buffer beam (replacing the existing arrangement of frames which came as far forward as the cylinders, then reached forward with separate extension frames) with its attendant modified cylinders and smokebox saddle, and providing a bogie with deep plate-frames instead of the open-framed pattern which had featured for so long on previous 4-6-0 classes, with the exception of the 'Kings'.

This re-designed front end now appeared on No 1000, along with a much larger boiler than had previously been used on two-cylinder 4-6-0s, and whose construction had made use of flanging plates which had been used for the boilers of wartime Swindon-built LMS '8F' 2-8-0s. Another departure from standard was the introduction of 6 ft 3 in. diameter driving wheels, topped overall by a long continuous splasher – a feature borrowed from the two streamliners, 5005 *Manorbier Castle* and 6014 *King Henry VII*, of the mid-1930s. Another detail borrowed from the bullet-nose era was the re-introduction of sliding ventilators in the cab roof (No 1000 was without them initially – it got them later), this roof now meeting the turn at the top of the cabside sheets almost flush, and without the familiar overhanging rain strip along the joint. An absence of brass trim was noticeable, too, particularly the familiar vertical brass beading on the front corners of the cab – brought about, no doubt, by the shortage of brass on that early post-war period of austerity. Another, and novel, feature of the 'Counties' was the introduction of a footstep attached to the front buffer beam, necessitated by the 6 in.-wider-than-usual cab, whose extra width allowed little or no toe-hold on the plates adjoining its bottom edges (hence no cabside grab-rail) and therefore access to the running plate

had to be gained via the front buffer beam.

The tender introduced with the 'Counties' was completely new. Gone were the traditional Great Western frames with their familiar 'three-a-side' angular drop-down sections embracing the axle-box horn cheeks (features since the very earliest days of the GWR) and instead, the frames now had full-depth side plates with half-moon cut-outs – a pattern widely used by other railways. A complete departure, too, was the all-welded flush-sided superstructure with a completely re-designed footplate area, embracing built-in lockers. Early tenders of this new design had an uninterrupted flush side from top to bottom but later ones appeared with an angle strip welded all around their bottom edges.

By the time the 'Counties' were introduced in 1945, the GWR was well on the way to restoring its engines to their pre-war finery, but never went back to lining-out below the footplate. The one exception was *County of Middlesex*, which was to have the distinction of being the only post-war Great Western engine to have orange lining on its underworks. Another unique distinction bestowed upon *County of Middlesex* was the double chimney, being the only 'County' to get one from new – the only Swindon-built engine in GWR days to be thus adorned.

With 280 psi boiler pressure (influenced, perhaps, by the Southern Railway, who had already inched up pressure to the same figure on their 'Merchant Navy' and 'West Country' classes) and 6 ft 3 in. driving wheels, the tractive effort produced (32,580 lb) warranted No 1000 to receive initially the power classification 'E' on its cabside red discs when new, but as subsequent engines appeared they were rated 'D – X' ('X' for Extra Availability) and *County of Middlesex* was similarly down-rated the following March when she appeared sporting nameplates, as well as rooftop sliding ventilators, as seen opposite.

Following the precedent of the first batch of an earlier class of 4-4-0 'Counties', the new engines took their names from English and Welsh counties – though one wonders why 1001 *County of Bucks* was so named, and the more correct (and much more lyrical) title 'County of Buckingham' not spelled out in full; it couldn't have been because of the post-war shortage of brass as its predecessor, No 3811, in the early part of the century had been *County of Bucks* too.

GWR No 1000 *County of Middlesex*

Talyllyn Railway No 3 SIR HAYDN

ABUNDANT in Wales are the narrow-gauge steam railways with their 'Great Little Trains of Wales' which have long since become established as tourist attractions and are renowned the world over. Whilst on a tour of Germany with his traction engine, the author met many Germans – not all steam enthusiasts – who admitted that they 'came to England for their holidays to enjoy the little trains of Wales' (!), for such is their fame. Each has its own charm, and none more so than the Talyllyn Railway which wanders from the little town of Towyn through $7\frac{1}{4}$ miles of beautiful Welsh scenery to Nant Gwernol. This railway was incorporated by Act of Parliament on 5 July 1865 and completed the following year.

Like others of these little railways, the 2 ft 3 in. gauge Talyllyn was laid originally to handle slate traffic, bringing supplies from the Bryn Eglwys Quarry in the hills down to the port town of Towyn for shipment or to link up with the main-line railway. The decline in the demand for slate brought disuse and dereliction to the little railways and they could have passed into oblivion had it not been for the efforts of one man, the late L. T. C. (Tom) Rolt, an engineer who had turned writer on railway matters and 'things mechanical'. Aware of the possibilities of the Talyllyn Railway, he called for volunteers and formed a preservation group in 1950 who rolled up their sleeves and, against all odds, eventually got what was to be Britain's first preserved railway in action, this time to carry passengers using an assortment of little engines.

One of these engines is Talyllyn Railway No 3 *Sir Haydn*, well over a hundred years old and still going strong, and named after Sir Henry Haydn Jones, a one-time owner of the railway. Built in 1878 as an 0–4–0 tank by the Falcon Engine Works at Loughborough, it worked for many years on another slate line, the Corris Railway. During its century of service, the little engine has been rebuilt twice – by its makers in 1900 and again by the Talyllyn Railway in 1968, when it was given its trailing truck and impressive extended cab, built by apprentices at Brush Electrical Machines of Loughborough, the successors to its original builders. Sporting a whistle from a defunct 'Britannia' class engine and a full-size locomotive headlamp, *Sir Haydn* is shown here in the dark green livery it wore after that second rebuild.

With its captivating charm, *Sir Haydn* was deliberately chosen to be a finale to this selection of locomotive portraits as a tribute to Tom Rolt and his enterprise which paved the way for so many of the restored railways operating steam locomotives in Britain today – not forgetting all those volunteers who lend a hand – be they 'full-size' or narrow-gauge.

Talyllyn Railway No 3 *Sir Haydn*

Technical Note: About the Illustrations

Like a number of boys with whom I shared my schooldays (and believing, no doubt, that I had a greater knowledge of locomotives than those engineers qualified to design them) I confidently pencilled my own ideas – more often than not on the backs of exercise books – of what a streamlined 'King' should really look like and designed gigantic twelve- and fourteen-coupled monstrosities with little regard for the limits of the loading gauge, and whose tractive efforts were far in excess of practicality and greater than anything ever likely to be required by any railway.

In the fullness of time, these doodles settled down to serious drawing, eventually to become illustrations such as those appearing in this book. Each one has started as a pencil line – the rail line – from which the outline of the image has been built up, mostly at a scale of $\frac{1}{2}$ in. to 1 ft (nothing metric), as much time being spent studying works' drawings and photographs – sometimes the real thing – as producing the outline with all its detail.

I had no training with the airbrush. In the Royal Navy I once had an old chief petty officer who used to say, 'Don't volunteer for anything, lad, but if you do, tell 'em you can do it, then go away and find out how to do it.' Applying his wisdom when, as a freelance with a young family to feed, I was offered a stack of technical photographs to retouch, I took them on, bought an airbrush and found I enjoyed the textures and tones it produced. Airbrushing is a long and painstaking business involving masking-off the individual areas to be coloured (this taking much longer than the airbrushing process itself) and, using watercolour, the locomotive gradually takes shape over a period of a month or six weeks, helped along with the finest of sable brushes to fill in tiny details such as rivets.

Locomotive colours and liveries have their own fascination for me. Those men who painted railway engines in distant days had only a few colours to call on compared with the myriad of hues in today's colour charts, and I know of only one old-established paint company whose range of colours is still limited to more or less the range available to those old locomotive painters. They were artists in their own right, using colour to complement shape and lining it out wisely and sympathetically. Great Western green, for example (*never* yellow-based Brunswick green, no matter how often repeated as being so by 'experts'), was middle chrome green, derived from chrome orange pigment, which is why its chrome orange lining 'sat' so nicely there in harmony, never offending or clashing, the two colours mellowing together over a period of time as they both sprang from the same source.

A number of engines, I am told, face the wrong way. This was brought home one winter's night in a cosy pub when a print of *Pendennis Castle* was under discussion. At the end of the evening, and with our host insisting *please* would we leave, we had decided that 'right/wrong way round' for an engine reflected an inborn association with horses, those generations of our forebears who had ridden armed with swords having mounted with their horse's head on the left, and which, one sage remarked, is why our native British galloping-horse fairground roundabouts revolve clockwise. It follows that it is more natural to view a locomotive with its smokebox to the left.

Food for thought, but until then I had never considered the matter, not being a particularly 'horsy' person. I have faced the engines whichever way my fancy took me, or in a way in which I wished to show some particular detail which was only on one side. The other side of *County of Middlesex* is rather plain and uninteresting.

As for the notes accompanying each illustration, I have tried not to be too technical, though, needs must, technicalities have crept in. Rather, I have dotted down such interesting details as have come to light as each locomotive has taken shape. Some of the illustrations were commissioned; some were simply self-indulgence. All were pleasurable to produce.